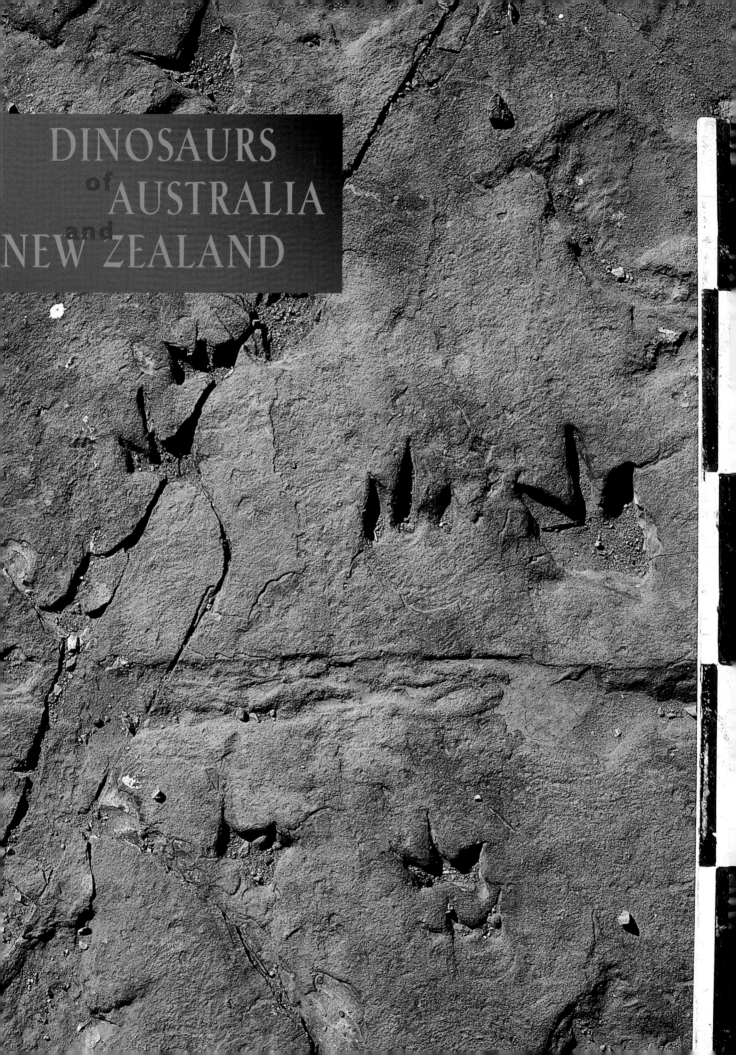

DINOSAURS of AUSTRALIA and NEW ZEALAND

DINOS

AUST

and

NEW ZE

and other animals

John A. Long

AURS
of
RALIA
ALAND

of the Mesozoic Era

Harvard University Press

Cambridge, Massachusetts

contents

For Sarah, Peter and Madeleine

WOOLUNGASAURUS GLENDOWERENSIS — NATIONAL PHILATELIC COLLECTION, AUSTRALIA POST. ILLUSTRATION BY PETER SCHOUTEN

Published by arrangement with
University of New South Wales Press Ltd
Sydney, Australia

Library of Congress Cataloging-in-Publication Data

Long, John A., 1957–
Dinosaurs of Australia and New Zealand and other animals of the Mesozoic era /
John A. Long.
p.   cm.
Includes bibliographical references and index.
ISBN 0-674-20767-X
1. Dinosaurs—Australia. 2. Dinosaurs—New Zealand. 3. Paleontology—Mesozoic.
4. Animals, Fossil—Australia. 5. Animals, Fossil—New Zealand.   I. Title.
QE862.D5L5792   1998
567.9'0994—dc21   97-41371

Designed by Di Quick and Dana Lundmark
Printed by South China Printing, Hong Kong

acknowledgements

Several people kindly read over and checked parts of the text, or contributed freely in discussion of the Australian and New Zealand dinosaur faunas, and have provided photographic materials or artwork for use in this book. My heartfelt thanks go out to the contributing artists, Mr Tony Windberg, Mr Brian Choo, Peter Schouten, Peter Trusler, Mike Skrepnick, Bill Stout and Ms Emily Dortch (Perth); and to the scientists—Dr Anne Warren (La Trobe University), Dr Ralph Molnar (Queensland Museum), Dr Pat Vickers-Rich (Monash University), Dr Tom Rich (Museum of Victoria), Dr Arthur Cruickshank (Leicester Museum and Art Gallery, UK), Dr Tony Thulborn and Mr Colin McHenry (University of Queensland), Dr Angela Milner (Natural History Museum, London), Mr Michael Shapiro (Harvard University), Dr Ewan Fordyce (Otago University, New Zealand), Dr Norton Hiller (Canterbury Museum, Christchurch, New Zealand), and Mrs Joan Wiffen (Hawkes Bay, New Zealand). Dr Chris McGowan (Royal Ontario Museum, Canada) also reviewed the manuscript and gave helpful comments.

In the Western Australian Museum, I acknowledge my colleagues for their many helpful conversations and guidance over the years: Dr Ken McNamara, Dr Alex Bevan and Dr Alex Baynes; Ms Kristine Brimmell for her excellent photography and field assistance; Mrs Jen Bevan for editorial assistance; and Mrs Danielle West for drafting and secretarial assistance.

The directors and trustees of the Western Australian Museum are also thanked for their support of the project since its beginning in 1990. Last, but not least, I sincerely thank the editorial team at UNSW Press for their assistance in producing this edition.

This book attempts to provide amateur and specialist alike with useful information on Mesozoic animals in the Australasian region, and in order to do this one treads a very fine line between including too much technical jargon and making the text interesting to uninitiated dinosaur fans. So, this book provides some information about each animal in terms of its discovery and the circumstances surrounding the finding of each fossil, some technical information about why each was named as a new species, where its name derives from, and how the animal is placed in relation to other members of the same group or family. Where possible, I have tried to illustrate all the most important specimens as they exist in museums around Australia and in New Zealand, and in some cases to show reconstructions of the animals or their close relatives. In some ways this book highlights the paucity of Australasian dinosaur material whilst showing the diversity of amphibian and other reptile life at the time of the dinosaurs. I hope that it contains something of interest for everyone—from the child doing a school project to the professional palaeontologist who wants an accurate compilation of Australia's and New Zealand's Mesozoic animals.

In the 1990s Australia and New Zealand are entering their most exciting time ever for dinosaur research. During this period we have seen the startling discovery of several groups of dinosaurs with a presence in Australasia earlier than recorded elsewhere: some of these were previously thought to be unique to the Northern Hemisphere (for example, possible ceratopsians, ornithomimosaurs and an oviraptorid).

The reconstructions of dinosaurs provided here are based on what is known of the Australian and New Zealand fossils together with knowledge of their nearest well-known relatives, and this is combined with a consideration of their possible environment to provide an idea for colour schemes. As all colour schemes for extinct animals that are based on bones are speculative, I make no apologies for the exotic colours or reconstructions I have presented. They are just how I, and the artists working with me, imagine the creatures might have looked.

To complete the picture of life in the world of dinosaurian Australia and New Zealand, I have included details of the ecosystems and plant life of those times, and present some of the more exciting finds in full colour. I hope it presents a full, more exciting picture of the ancient Australasian region and will serve well for years to come as a basic reference useful to the general public and students of palaeontology alike. The reader who wants to find out more about the many kinds of fishes living in the seas, rivers and lakes of these periods can refer to my previous book, *The Rise of Fishes: 500 Million Years of Evolution* (UNSW Press, 1995/JHUP, 1995).

Kronosaurus, a giant marine reptile from Queensland

part one

theST

DISCO

of DIN

UDY

and

VERY

OSAURS

one

DINOSAURS AS FOSSILS

The word 'dinosaur' is created from the Greek words 'deinos', meaning 'terrible', and 'sauros', meaning 'lizard'. It was first coined by the famous English anatomist and palaeontologist, Sir Richard Owen, in 1841 at Plymouth, England, to describe a group of large, extinct reptilian animals. Despite the reference to lizards in the original meaning of their name, dinosaurs are actually more advanced animals, in the evolutionary sense, than modern lizards or other reptiles: dinosaurs evolved well after these other groups had first appeared, and would be better considered as representing creatures at the peak of reptilian evolution. Most scientists now accept the notion that all living birds are descendants of small, agile carnivorous dinosaurs. Therefore, the dinosaurs did not completely die out some 66 million years ago as we were all taught to believe! Dinosaurs should be regarded, then, as resembling birds more than they do most of the living reptile groups, such as turtles and lizards; crocodiles, however, are slightly closer to dinosaurs than the other living reptile groups.

Since the days of Sir Richard Owen, when only seven dinosaur species were known (in 1841), our knowledge of this fascinating group has greatly increased as a result of thousands of new finds of dinosaur remains from every part of the globe. Over 500 species of dinosaurs belonging to some 360 genera have now been named, and at least ten new forms have come to light each year since the 1980s. Dinosaurs are now a part of our culture and represent for many children their first introduction to the world of science. Following the release of the popular movie *Jurassic Park* in 1993, dinosaurs became a billion-dollar-a-year industry. Humans have had a relatively short time on this planet (less than 4 million years as a genus, and about 100 000 years as an advanced species) compared to dinosaurs, which ruled the Earth for over 150 million years. This makes dinosaurs the most successful group of backboned animals ever to have lived on land. Humankind still has a long, long way to go even to approach this record.

The better-known dinosaurs mostly come from sites in the Northern Hemisphere. Our knowledge of the group is based principally on major sites in North America, central Asia, Europe, and Africa. However, knowledge of South American and African dinosaurs has been greatly increasing over recent years, as it has for dinosaurs from Australia and Antarctica. Many recent discoveries from the 1980s in Australia and New Zealand are beginning to fill in the picture of what types of dinosaurs, and other creatures of their time, used to live in the Australasian region. What did they look like? What did they eat? How did some dinosaurs survive in cold climates in some parts of Australasia?

This book will try to answer these and other questions and explore the world of Australian dinosaurs—from early discoveries to recent finds, from the enigmatic pieces to the nearly complete skeletons, and from sites as far afield from each other as the northern opal fields of South Australia and New South Wales, the blustery coastline of western Victoria, the hot outback deserts of central Queensland and the cool, forested streams of New Zealand.

HOW TO USE THIS BOOK

It may be necessary for some readers to familiarise themselves with the background information that scientists take for granted when discussing dinosaurs—such as what fossils are, how old they are, and how they are dated; what the planet was like when dinosaurs lived, and why they died out. This introductory chapter deals with such background information. The biological background to dinosaurs, their evolutionary origins, their family tree, and details of their general anatomy are given in chapter 2.

In the main chapters of the book a brief introduction is provided to the environmental settings in the Australasian region for each geological time period of Australia and New Zealand. In the listings of faunas for each period the entries have been designed so that for each animal there is, first, a general information section (for use by the non-specialist); this is followed by a technical data section (aimed specifically for academic use), which defines the characteristics of each species and, unavoidably, requires the frequent use of technical terms. There is also a glossary of scientific terms at the end of the book, which provides explanations for some of the more technical language used in discussing the characteristics of the fossils and their environments.

Wherever possible, representative illustrations of actual fossil specimens are included. Some imaginative reconstructions of the animals are also shown, as guidelines to how the creature may have looked when alive, based on reasonable scientific estimates from what limited data is provided by the actual fossils.

FOSSILS

Fossils are the remains, impressions or trackways made by living organisms in past time. The word 'fossil' comes from the Latin *fodere* (meaning 'to dig'), or, more specifically, from *fossilis*, meaning something that was 'dug up'. The study of fossils is called 'palaeontology' (from the Greek, meaning 'study of ancient life'), and 'palaeontologists' are scientists who carry out research on fossils. The aim of such research is not confined to reconstructing the past life of this planet; it also involves determining the nature of past environments, the ages of certain rock layers (which may point the way to new mineral or petroleum discoveries) or observing and describing the patterns of evolutionary change in lineages of organisms. Thus, the study of fossils can help us understand why certain communities of animals and plants exist together, today, in often complex relationships. Most importantly to human nature, the study of fossils gives us some insight into where we came from, and where we are heading as a species.

Common fossils that might be found are bones of extinct animals, or impressions of plant leaves or petrified wood; but impressions left by the feet of dinosaurs are common, too, as are those of the

in fact is still present; it is the small spaces in the bone that have been filled in with minerals crystallised out of groundwater, which has coloured the bone and altered its overall composition. This process is known as either calcification, when calcite is deposited between bone pore spaces and cavities, or as phosphatisation, when phosphatic minerals impregnate the bone. There may also be silification or opalisation.

In the majority of cases bones are not fossilised. Instead, they are eaten away by scavengers, or decay into dust through regular processes of weathering, abrasion and bacterial action. It is only in those rare cases when an organism is rapidly buried after death—such as in a flood, or by floating out to sea and sinking in quiet, muddy conditions—that its remains may become entrapped within layers of sediment, to later

▷ Skeleton of *Camarasaurus*, a plant-eating dinosaur, 17 m long, that lived 150 million years ago in North America

BASIL BALME

feeding burrows made by worms, and the microscopic grains of pollen that occur in ancient sediments. The bones of a dead animal may become impregnated with minerals during the long period of time they are buried underground, and eventually take on a petrified appearance, with a darker coloration and much greater weight than a modern animal bone has. This is why people often refer to dinosaur bones as 'turned to stone' or 'petrified'. However, the original bone

become impregnated with additional minerals and thus be preserved as fossil bones. Then, further chance comes into play—as to whether the layer of sedimentary rock (which is called a 'horizon') that contains the fossil bones will ever be exposed again at the Earth's surface, weathered away sufficiently for the fossil bones to be revealed. Still further luck is then required for any person to stumble across those layers at the precise time when the bones are exposed, and recognise them

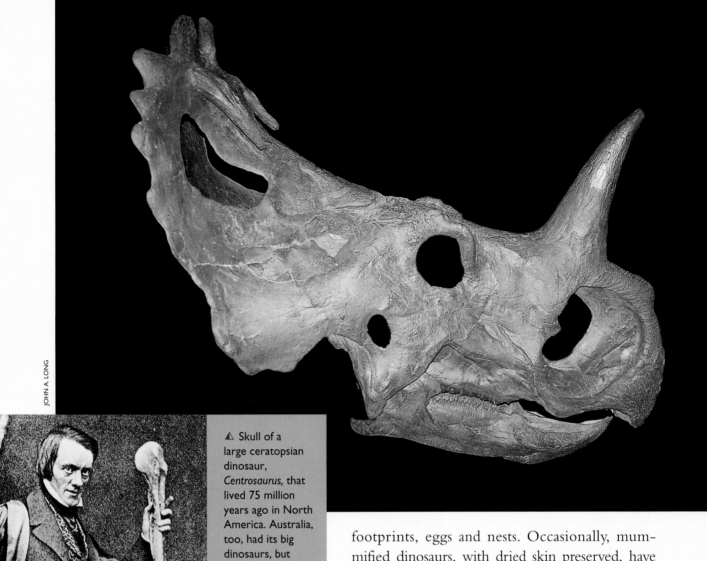

JOHN A. LONG

△ Skull of a large ceratopsian dinosaur, *Centrosaurus*, that lived 75 million years ago in North America. Australia, too, had its big dinosaurs, but most are known only from scant remains

◁ Sir Richard Owen, who first coined the word 'dinosaur' in 1841 to describe a group of large, extinct reptiles

as fossil bones. Nowadays most dinosaurs are found by experts who know where to look for them and how to recognise their remains from small fragments exposed on the surface of weathering cliffs or erosion gullies.

What is known of dinosaurs today has largely been learned from the study of their isolated bones, or sometimes from partial (or in very rare cases, complete) skeletons, but information about them can also be obtained from their fossilised footprints, eggs and nests. Occasionally, mummified dinosaurs, with dried skin preserved, have been found; although it is more common in recent years for impressions of dinosaur skin and dermal scales to be identified. Australia's and New Zealand's dinosaurs are represented mostly by isolated and fragmentary bones; but there are some rare, partially articulated remains, including nearly complete skeletons of two dinosaurs, and even the skin denticles of one armoured dinosaur. Many fossilised dinosaur footprints have also been found throughout Australia. In order to study the environment and climate that dinosaurs lived in, not only do all the fossils found with the dinosaur have to be considered, but also the data held in the geological record of the rocks. By comparing the ancient plant assemblages with their modern counterparts, and by studying the geological record of climatic indicators (such as the presence of ancient glacial deposits or tropical coal swamps, and so on), the ancient ecosystems that dinosaurs inhabited can be reconstructed.

GEOLOGICAL TIME

Geologists gauge the relevant changes in the movements of continents, the formation of geological structures and the evolution of life on Earth in terms of millions of years—in contrast to the lifespan of a single person, which might be 100 years if one is really lucky. The age of the Earth has been calculated at approximately 4 600 000 000 years, or 46 x 10⁶ long human lifespans! The oldest minerals in the world are from Mt Jack, near Newman, Western Australia. An ancient conglomerate rock from Mt Jack contains small crystals of the mineral zircon that have been dated as being up to 4.3 billion years old—that is, dating from the time just after the Earth's primitive crust had cooled long enough for the first rocks to be formed.

Mountain chains form and their rocks weather away through erosion in a continual cycle that causes layers of sedimentary rocks to be deposited in broad topographic depressions known as sedimentary basins. An example of this process is the Mississippi Basin in North America: there, for over 140 million years sediments carried down the Mississippi River have been continuously deposited into the Gulf of Mexico, as the edge of the continent slowly sinks, which allows a further inflow of sediments carried by the waters of the Mississippi River. The build-up of layers of sedimentary rocks over the long history of the Earth has resulted in many different horizons of sedimentary rocks, some of which are now exposed in various gullies, canyons and coastlines around the world.

The first geologist to recognise that successive layers of sedimentary rocks contained different assemblages of fossils

was an Englishman, William Smith, who made this discovery in the early 1800s. Since then other scientists, from recognising different assemblages of fossils, have built a geological time-scale which chronologically frames the sequence of life-forms from simple algae to complex vertebrates (backboned animals) such as humans.

Each time horizon that is characterised by a certain type of life is called a 'Period'—for example, the Devonian Period. These periods are grouped in three main eras: the Palaeozoic ('ancient life'), Mesozoic ('middle life') and the Cenozoic ('new life'). These three eras collectively form the Phanerozoic Eon ('the age of life'). Before the Phanerozoic was Precambrian time (4500–540 million years ago), in which little life existed on Earth apart from bacteria and algae and, towards the end of this time, some lower invertebrates. The Precambrian is divided into the Archaean Eon (4500–2500 million years ago),

JOHN A LONG

and the Proterozoic Eon (2500–540 million years ago). In the Archaean the only fossils are bacteria and algae, sometimes forming mound structures called stromatolites. In the Proterozoic Eon (which means 'first life') abundant microfossils of single-celled organisms occur, including the first cells with a nucleus (eukaryotes), as well as multi-celled animal life (metazoans) such as jellyfish, worms and coral-like forms. A famous site in the Flinders Ranges of South Australia, Ediacara, shows one of the best assemblages of well-preserved late-Proterozoic soft-bodied fossils in the world.

Rocks and fossils are dated by two main methods: radiometric dating, and relative age dating. Radiometric dating measures the actual decay of a radioactive isotope within a mineral in an igneous rock; it uses the knowledge, that the crystals of the mineral formed at the time the molten rock cooled. From this stage onwards the radioactive isotopes of certain elements began to decay to other elements or isotopes. By calculating the decay rate of these isotopes, and applying this rate to the proportions of decayed elements in the mineral, it is possible to derive a fairly accurate age for a rock. Radiometric dating uses the ratios between several different isotopes and their known decay rate, for example, between potassium and argon, or rubidium and strontium; or, for geologically younger items, carbon 12 to carbon 14 isotope ratios. Thus, an age obtained on a rock horizon can then be related to the fossil-bearing horizons above or below it. The second method of estimating the age of a fossil—in other words, relative age dating—is based on the fact that there is a large number of known, dated horizons around the world. Because of this it is now possible to tie in almost any horizon containing fossils to a dated horizon within the local region, or to other

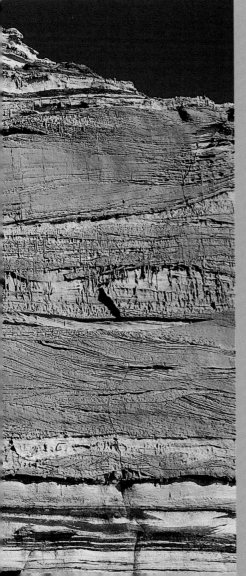

◁ Layers of sedimentary rocks exposed at the surface of the Earth often contain fossils. These cliffs, at Kalbarri in Western Australia, represent ancient sand dunes

TODAY

weathering exposes layers with fossils

bone

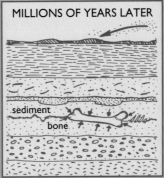

MILLIONS OF YEARS LATER

sediment

bone

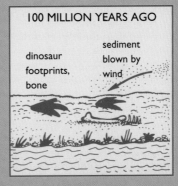

100 MILLION YEARS AGO

dinosaur footprints, bone

sediment blown by wind

◁ How a fossil is formed. A dinosaur has died near a river (bottom), and the carcass is preyed upon by scavengers. Sediment carried by the wind is blown onto the footprints. Eventually the bones are buried by flood waters, washed into the main river channel and covered with sands carried along by the currents. Footprints are sometimes preserved if covered by wind-blown sands. After millions of years of continual deposition of river sediments, the bones are buried well below the surface (centre) and start to become fossilised through interaction with mineral-rich groundwaters. Millions of years pass and the rock layers are pushed to the surface by movements in the Earth's crust. Erosion occurs, eventually exposing again the layers of sands (now turned to sandstone) containing the fossils (top)

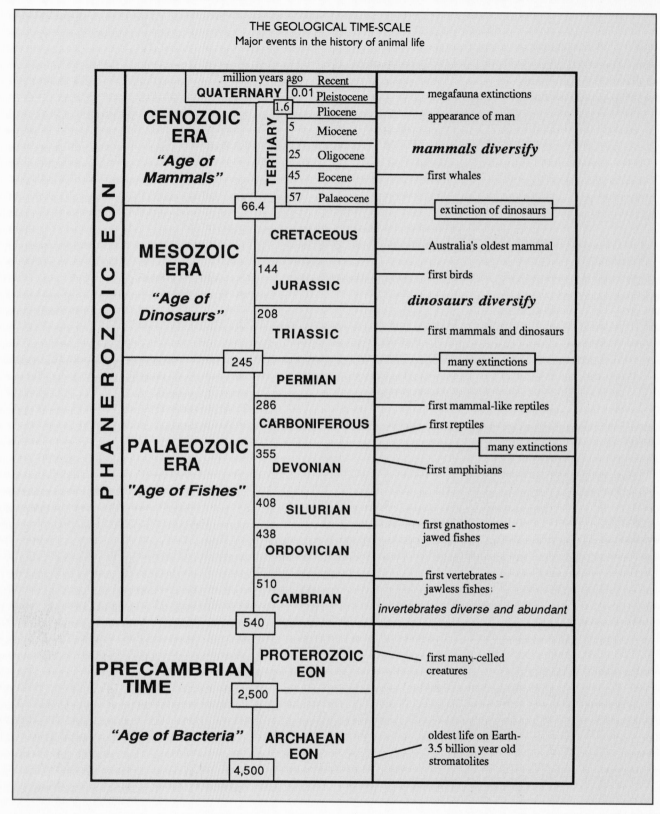

THE GEOLOGICAL TIME-SCALE
Major events in the history of animal life

			million years ago		Major events
PHANEROZOIC EON	**CENOZOIC ERA** *"Age of Mammals"*	**QUATERNARY**	0.01	Recent	megafauna extinctions
				Pleistocene	
			1.6	Pliocene	appearance of man
		TERTIARY	5	Miocene	*mammals diversify*
			25	Oligocene	
			45	Eocene	first whales
			57	Palaeocene	
			66.4		extinction of dinosaurs
	MESOZOIC ERA *"Age of Dinosaurs"*	**CRETACEOUS**			Australia's oldest mammal
			144		first birds
		JURASSIC			*dinosaurs diversify*
			208		
		TRIASSIC			first mammals and dinosaurs
			245		many extinctions
	PALAEOZOIC ERA *"Age of Fishes"*	**PERMIAN**			
			286		first mammal-like reptiles
		CARBONIFEROUS			first reptiles
			355		many extinctions
		DEVONIAN			first amphibians
			408		
		SILURIAN			first gnathostomes - jawed fishes
			438		
		ORDOVICIAN			
			510		first vertebrates - jawless fishes;
		CAMBRIAN			*invertebrates diverse and abundant*
			540		
PRECAMBRIAN TIME		**PROTEROZOIC EON**			first many-celled creatures
			2,500		
	"Age of Bacteria"	**ARCHAEAN EON**			oldest life on Earth- 3.5 billion year old stromatolites
			4,500		

horizons containing a similar fossil assemblage known elsewhere to be near a dated rock sample.

In this way the geological time-scale has been built up and refined, not only through recent advances in radiometric dating technology, but also as a result of an increased precision in describing and comparing new finds of fossil assemblages. Consequently, when a new fossil is discovered, it is often not very hard to put an age on the specimen either by referring to the known assemblage of other fossils in the rock (that is,

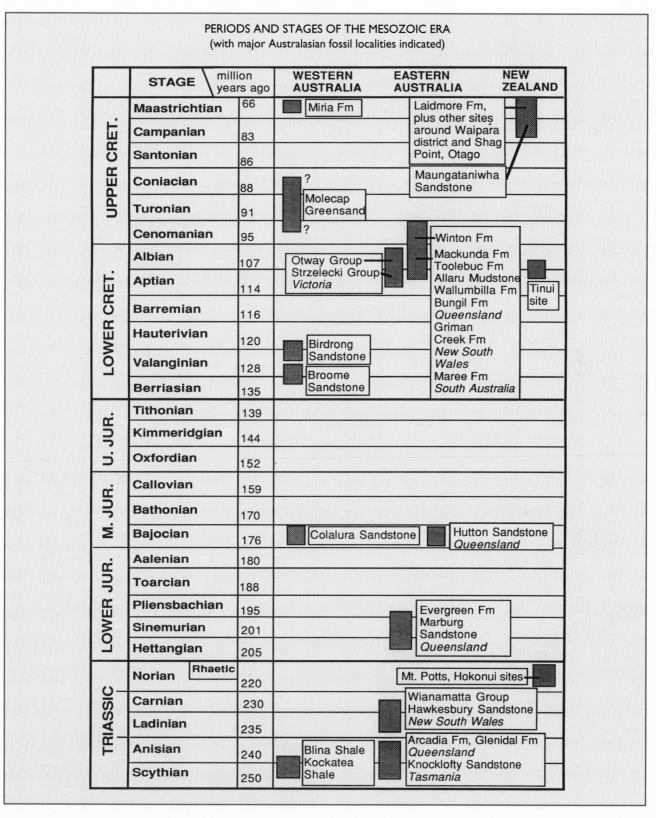

PERIODS AND STAGES OF THE MESOZOIC ERA
(with major Australasian fossil localities indicated)

	STAGE	million years ago	WESTERN AUSTRALIA	EASTERN AUSTRALIA	NEW ZEALAND
UPPER CRET.	Maastrichtian	66	Miria Fm	Laidmore Fm, plus other sites around Waipara district and Shag Point, Otago	
UPPER CRET.	Campanian	83			
UPPER CRET.	Santonian	86			
UPPER CRET.	Coniacian	88	?	Maungataniwha Sandstone	
UPPER CRET.	Turonian	91	Molecap Greensand		
UPPER CRET.	Cenomanian	95	?	Winton Fm	
LOWER CRET.	Albian	107	Otway Group, Strzelecki Group *Victoria*	Mackunda Fm, Toolebuc Fm, Allaru Mudstone, Wallumbilla Fm, Bungil Fm *Queensland*, Griman Creek Fm *New South Wales*, Maree Fm *South Australia*	Tinui site
LOWER CRET.	Aptian	114			
LOWER CRET.	Barremian	116			
LOWER CRET.	Hauterivian	120	Birdrong Sandstone		
LOWER CRET.	Valanginian	128	Broome Sandstone		
LOWER CRET.	Berriasian	135			
U. JUR.	Tithonian	139			
U. JUR.	Kimmeridgian	144			
U. JUR.	Oxfordian	152			
M. JUR.	Callovian	159			
M. JUR.	Bathonian	170			
M. JUR.	Bajocian	176	Colalura Sandstone	Hutton Sandstone *Queensland*	
LOWER JUR.	Aalenian	180			
LOWER JUR.	Toarcian	188			
LOWER JUR.	Pliensbachian	195		Evergreen Fm, Marburg Sandstone *Queensland*	
LOWER JUR.	Sinemurian	201			
LOWER JUR.	Hettangian	205			
TRIASSIC	Norian / Rhaetic	220		Mt. Potts, Hokonui sites	
TRIASSIC	Carnian	230		Wianamatta Group, Hawkesbury Sandstone *New South Wales*	
TRIASSIC	Ladinian	235			
TRIASSIC	Anisian	240	Blina Shale, Kockatea Shale	Arcadia Fm, Glenidal Fm *Queensland*, Knocklofty Sandstone *Tasmania*	
TRIASSIC	Scythian	250			

its associated fauna or flora), or by isotopic dating of volcanic rocks above and below the fossil layer to provide the parameters of its age range. The science of correlating and dating rock formations using fossil assemblages is called biostratigraphy.

Microfossils are the microscopic remains of animals or plants, which often occur in great abundance in small amounts of marine sediments. Thus, a spoonful of mud from the ocean floor may contain hundreds of microscopic fossils of protozoan animals such as the single-celled

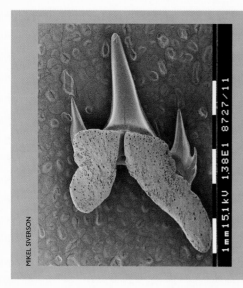

MIKEL SIVERSON

▷ A foraminiferan fossil,
Hedbergella

◁ A fossilised shark's tooth,
Johnlongia.

Microscopic remains of
single-celled organisms
like foraminiferans enable
scientists to accurately
estimate the age of rocks
because the shapes of these
fossils changed rapidly
through time. Fossilised
shark's teeth are also useful
for dating sediments

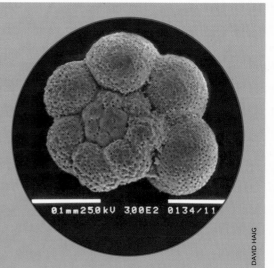

DAVID HAIG

foraminiferans or radiolarians, or many grains of plant pollen. Such fossils are known from the silica or calcium carbonate shell or hard organic wall built around the soft parts of the organism or spore. Microfossils are the principal tool used today for dating drill cores of sediments in the search for petroleum or gas deposits.

DRIFTING CONTINENTS

As a key to understanding the past life on Earth it is important to have some knowledge of what is called plate tectonics. In other words, it is necessary to realise that the Earth's crustal plates—and therefore the continents above them—have been in a continual state of movement. The major continents, as we know them now, sit on plates of continental crust up to 150 km thick and are surrounded by seas on oceanic crust only about 5 km thick. These plates move around on the Earth's surface by friction, possibly due to convection currents in the hot molten mantle below the crust. When plates meet, one plate may ride over the top of the other one, pushing the plate underneath it down into the Earth's hot mantle. This process is called subduction and it causes melting of rock in the subducted plate and may result in the formation of volcanoes as lighter density magma (molten rock) rises up to the surface. That is why most of the world's active volcanoes are found near where tectonic plates meet, such as around the Pacific rim, or along the margins of the

Southeast Asian and Australian plates. Similarly, major earthquake zones are also found in these regions as plates move against each other, or pieces of crust move along fault lines close to plate margins, and built-up pressures are released as sudden shocks.

From the evidence of palaeomagnetism (the study of the Earth's magnetic field trapped in rocks) scientists can determine which way the continents were orientated, with respect to the north and south poles, at specific geological times. This information, together with data on past climates from the rocks and fossils—for instance, the presence of glacial deposits or tropical limestone reefs—can give us a fairly accurate picture of where each of the continental plates was throughout past times. This concept is important to understanding the distribution in

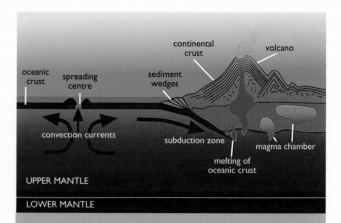

Subduction of the Earth's plates happens when one crustal plate overrides another, pushing it down into the mantle. This results in the formation of mountains and volcanoes

space and time of animals such as dinosaurs, since some groups died out in one part of the world, while continuing to survive elsewhere in isolation. In Australia, for example, there is a theory that the marsupials have radiated and become the dominant mammal group simply because Australia became isolated from Antarctica at a time when marsupials were a dominant mammalian group: so, marsupials flourished because other animals could not invade Australia and create competition.

Continental reconstructions given for each period of the Mesozoic Era show the movement of the major continents during the time when

The widespread distribution of many dinosaurs throughout the world occurred because of this land connection. By the time dinosaurs became extinct—at the close of the Cretaceous Period, some 66 million years ago—Gondwana had split up, leaving a pattern of continents similar to that which exists now, but in different relative positions. Australia and Antarctica were the last continents to split at this time, thereby finally breaking up Gondwana.

Dinosaurs entered Australia and New Zealand through this connection with Antarctica, and then many of the groups evolved in isolation following the subsequent break-up of Gondwana. Only in

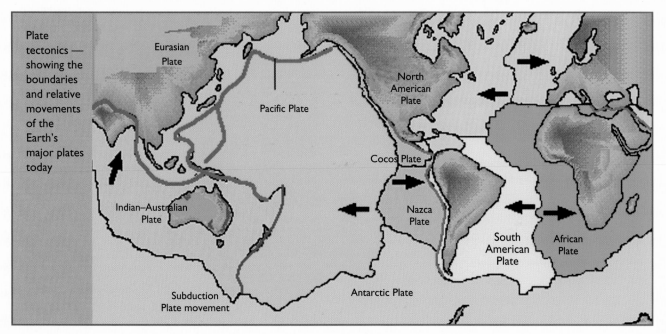

Plate tectonics — showing the boundaries and relative movements of the Earth's major plates today

Eurasian Plate

North American Plate

Pacific Plate

Cocos Plate

Indian–Australian Plate

Nazca Plate

South American Plate

African Plate

Subduction Plate movement

Antarctic Plate

dinosaurs were the dominant vertebrate group on the land. It is important to note that at that time Australia and New Zealand were part of a much larger supercontinent, Gondwana, which included Antarctica, India, South America and Africa, as well as other smaller crustal blocks. Discoveries of the little aquatic reptile *Mesosaurus* in Late Permian rocks of South America, and South Africa gave early indications that these continents may have once been joined together—a view later reinforced by many other similar fossil finds unique to these lands. Early in the Mesozoic Era Gondwana joined the northern supercontinent, Laurasia (comprising North America, Europe and Asia), to form a giant landmass called Pangaea.

recent years has the full picture been emerging as new discoveries are made both in Australia and Antarctica. In the 1990s a large carnivorous dinosaur was found in Lower Jurassic sediments on Mt Kirkpatrick, in the Transantarctic mountains, and named *Cryolophosaurus ellioti* (Hammer and Hickerson 1994). Associated with its skeleton were the remains of other dinosaurs (prosauropods) and a mammal-like reptile. This find, together with the discovery of other dinosaurs in western Antarctica, provides the evidence that Antarctica then had a thriving community of animals and plants, and that it was a gateway for entry of dinosaurs and other land animals into Australia and New Zealand.

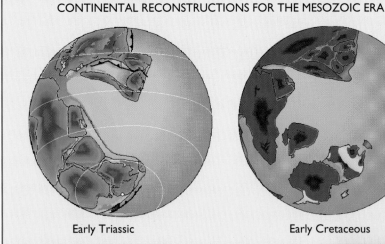

CONTINENTAL RECONSTRUCTIONS FOR THE MESOZOIC ERA

Early Triassic   Early Cretaceous

DISCOVERY AND PREPARATION OF FOSSILS

One of the questions palaeontologists are commonly asked is how they knew where to go and find a specific fossil. Such discoveries are not due to a magical sixth sense possessed by palaeontologists; they are often the result of much hard work looking for clues in public collections of fossils. In many cases a bone that has weathered out of a rock is found by an exploration geologist or a farmer, or anyone who just happened to be at some remote locality, and is brought into a museum for identification. The palaeontologist recognises it as a fossil and returns to the site to carry out further examination of the area for more bones, specifically looking for the particular layer of sedimentary rock from where the bones may have come.

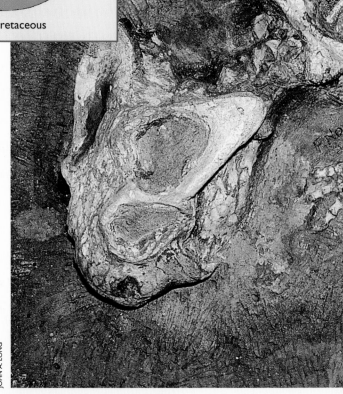

JOHN A. LONG

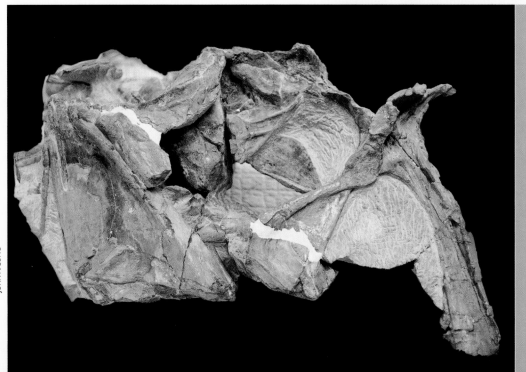

JOHN A. LONG

▲ Mammal-like reptiles from the Triassic such as *Diictodon* from South Africa provided early evidence suggesting the linking of the Gondwana continents

◁ The skull of *Cryolophosaurus ellioti*, an Early Jurassic theropod dinosaur, with a crest above its eyes, from Mt Kirkpatrick, Antarctica. Dinosaurs entered the Australasian region through Antarctica when the southern continents were joined as part of Gondwana

Sometimes geologists mapping a region will record the presence of bone fragments or collect samples of fossils to keep in the State Geological Survey collections. A palaeontologist searching in these collections, or in the vast collections of museums and universities, or those owned by private fossil collectors, may stumble upon a fossil not previously recognised by colleagues as being of scientific importance. The palaeontologist will chase up the lead using the locality information given on the specimen's label, which is not always as simple as it sounds. I have been to sites searching for fossils found in the early 1930s and have had to take many trips before locating the exact rock layers in which the fossils occur.

Another method of finding fossil sites is by looking at geological maps that indicate where rocks of the right type and the right age to contain fossils crop out in the field, and then going out to examine in detail these exposures of rocks (called 'outcrops'). Although often unproductive, this method may produce spectacular results as completely new fossil sites may be discovered with fossils never before seen by anyone, such as the Dinosaur Cove site, in Victoria. Nowadays satellite pictures enhanced with remote sensing technology are sometimes used to pinpoint likely fossil sites.

Fossil vertebrate sites can be divided into those which are readily productive and those requiring much hard work for little fossil yield. The first category includes sites which are most easily found, and often have been intensively worked by early palaeontologists. At such sites bones can be found simply by excavating a bone-rich layer of rock. In some cases, where the sites have been buried, they can be opened up and worked again if the proper earth-moving equipment is available to remove the overburden covering the deposit. In other sites there may be few fossils, but they are often of great scientific value, so the many hours of hard work put in for a few fragmentary bones is worthwhile. In these cases a site may consist of rock exposed over vast areas of land, requiring the palaeontologist's team to explore on foot every rocky outcrop or cliff, visually searching for a fragment of bone poking out. Weathering plays an important role in the working of such sites as they can be reinvestigated after several years and found to be productive again if the erosion rates have been high. Other sites require large volumes of rock to be excavated or washed down with sieves, followed by meticulous splitting or picking through the residues.

Rich fossil sites containing a single layer full of bones may be termed a 'bonebed'. Each piece of the bonebed is carefully excavated and split to reveal the fossils. If the material is limestone (a rock composed largely of calcium carbonate), large chunks can be transported back to the laboratory for special treatment. The bone can be completely freed from the rock in a process called 'preparing out'. This is done by dissolving the limestone in weak acetic acid; the bone is not dissolved because it is composed of calcium phosphate. A classic example of this type of preservation is seen in the treatment of the Gogo fish deposits of northwestern Australia, where 370-million-year-old fish skeletons have been prepared out in perfect three-dimensional form. Some of the Queensland dinosaurs, pterosaurs and marine reptiles have been found in limestones and can be treated in the same manner. The use of acids is a very slow process, however, and the bones must be continually hardened with plastic-based glues. Large bones must be prepared carefully in the field by being

▷ ▽ The little aquatic reptile *Mesosaurus* is found in Late Permian rocks in South Africa and South America. It gave early indications that these distant lands may have once been joined together as part of the ancient supercontinent of Gondwana

JOHN A. LONG

JOHN A. LONG

JOHN A. LONG

▲ A landscape showing potential for the discovery of dinosaur bones. Here sedimentary rocks of Early Cretaceous age are exposed over a large area of remote countryside in Western Australia

▷ A fish that is 370 million years old, *Onychodus*, from Gogo, Western Australia. This illustrates the beautiful preservation obtained from using acetic acid to prepare fossils out of limestone

▽ Excavating a fossil site in Western Australia where a pliosaurid skeleton was found

JOHN A. LONG

soaked with glues to internally harden the exposed bone. The rock also has to be undercut away from the fossil before the slab containing the bones can be covered with plaster jacketing to enable it to be safely transported back to the museum, where it undergoes further preparation with pneumatic airscribes or dental drills to remove the rock from the bones.

Once the fossils are in the museum, and if they are suitably resilient, there are several ways to prepare them out of the rock. Manual preparation by dental drills and small picks and chisels is the most common approach, especially when the surrounding sedimentary rock is softer than the mineralised bone. Sometimes, where the bones are still partially or fully articulated (joined as in life position), it is often preferred to leave the skeleton visibly exposed but still attached to the

KRIS BRIMMELL, WESTERN AUSTRALIAN MUSEUM

slab of rock in order to show these natural bone positions. If the bones are strongly mineralised, and can be freed from the rock without damage occurring, they can later be mounted in a reconstructed skeleton using metal framework support-structures. Chemical treatments to extract bones from rock are now becoming popular, as new methods have been developed using a wider range of acids—for example, acetic or formic acid for standard limestone dissolution, or thyloglycolic acid for disaggregation of iron-cemented sand-stones. In some cases where the bone is soft and crumbly, or mostly weathered away, and the surrounding rock is very hard, the bone itself is dissolved away using hydrochloric acid solution. The rock is then thoroughly washed in running water so that the hole where the bone once existed can be cast with latex or silicon rubber to produce a perfect replica of the shape of the bone.

RECONSTRUCTION OF DINOSAURS

A dinosaur can be reconstructed from only scraps of bone, or just from one small bone, such as was done with the Victorian *Allosaurus*—although it should be pointed out emphatically that unless more or less complete skeletal remains are known, any reconstruction based on fragments is at best only scientific guesswork. The method of reconstruction is to closely compare the anatomical features of the available bone with bones of other well-known dinosaurs and find which dinosaur, known from a relatively complete skeleton, has the most similar equivalent bone. If the similarities are within the known range of variation for a certain animal, it is then possible to build up a picture—from the single bone—of that whole animal, which is based on the known complete skeleton of the animal it most closely resembles. Particular characteristics that the single bone or bone fragment has can then be added to modify the reconstruction. For example, if the bone is more robust or more slender than that of the closest comparable relative, the reconstruction can be slightly modified to make the whole animal more robust or more slender. As well, to reflect a specific environment or climate the animal was living in, the reconstruction can be modified to incorporate features of an animal living in a comparable modern climate, based on the colour schemes and adaptations of animals living in such conditions today.

When a whole skeleton is preserved, the job of reconstruction is relatively simple. A good study of the anatomical features of the bones tells us, by comparison with modern animals, where muscles and ligaments were most likely attached—although caution must be exercised at this stage as their location is not always clear, and, even if correct, such a comparison does not always give a clear indication of the mass of those muscles. However, if biomechanical studies have been carried out on the bones to determine their stress

parameters, an estimation of muscle size can be reached that accords with the implied function of the bones. Such studies enable us to 'flesh the animal out' by adding a good estimation of the shape of the soft tissues on to the framework of the reconstructed skeleton. The skin is then added and a colour scheme chosen to match the animal's environment and probable lifestyle—for example, a small plant-eating dinosaur that lived in a lush tropical environment might require camouflage similar to the vegetation surrounding it to avoid being seen by predators. Using such a rationale, and from observation of all kinds of animals in today's forests and deserts, scientists can gain a good idea of what sorts of colour schemes might have adorned different dinosaurs.

From impressions preserved in the entombing sediments, dinosaur skin is now known from many kinds of dinosaur groups. The large theropods often had rows of enlarged dermal denticles; smaller theropods may have had feathers, a conclusion based on the 1996 discovery from China of *Sinosauropteryx*. Thus, combining all the different sources of information, it is possible to come up with reasonably accurate reconstructions of how dinosaurs may have looked in life.

◁ Palaeontologist at work preparing dinosaur bones. It is painstaking work, slowly removing the rock and hardening the ancient fossil bones

GREG WALLACE, WESTERN AUSTRALIAN MUSEUM

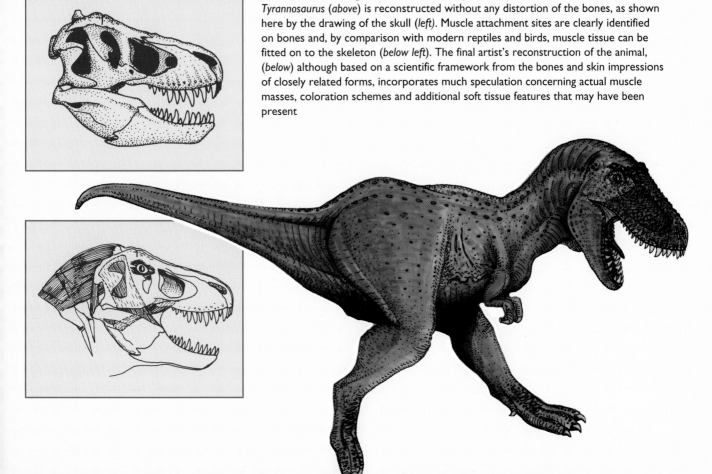

⊿ ▽ Reconstructing dinosaur soft anatomy from the bones. First, the skull of the *Tyrannosaurus* (*above*) is reconstructed without any distortion of the bones, as shown here by the drawing of the skull (*left*). Muscle attachment sites are clearly identified on bones and, by comparison with modern reptiles and birds, muscle tissue can be fitted on to the skeleton (*below left*). The final artist's reconstruction of the animal, (*below*) although based on a scientific framework from the bones and skin impressions of closely related forms, incorporates much speculation concerning actual muscle masses, coloration schemes and additional soft tissue features that may have been present

two

THE
FAMILY
TREE
OF
DINOSAURS

SOME BIOLOGY BASICS

Throughout this book animals are referred to as species or genera. Species are members of an animal or plant population which can interbreed, such as humans (of different races), or dogs (each species having many varieties). Different species include groups of animals which have morphological differences, and may not interbreed because of differences in habitat or because they lived at different times. Similar groups of species belong in the same genus. Within the genus *Homo*, to which we humans belong, the species *Homo sapiens* (modern man, us) could not interbreed with *Homo habilis*, as *H. habilis* lived about 4 million years ago. *H. habilis* differs in many anatomical features from us, making recognition from bones easy. Groups of genera sharing unique characteristics are classified into families, and families that share unique features with each other are placed into higher classificatory levels (superfamilies, infraorders, orders, subclasses, classes, and so on). This system is based on the work of the early Swedish naturalist Carolus Linnaeus and is called Linnaean classification.

This book deals with vertebrates, or back-boned animals, and the basic means of studying the fossils of such animals is from their skeleton. The skeleton of most higher vertebrates (excluding fishes) follows basically the same pattern of bones, with special modifications in each lineage or evolutionary group. The skull, which is composed of many bones, contains the brain and the main sensory organs and glands, as well as acting as an attachment site for neck and face muscles. The skull is important for fossil identification as it shows much variation between species and the teeth indicate the nature of the animal's diet. The arms and legs contain slender limb bones and hands (manus) and feet (pes), which may be modified as flippers in water-dwelling forms. Limbs tell us much about how an animal moved: its weight and proportions are often directly related to the shape of its limb bones, and its hands and feet are related to movement, feeding habits, defence strategies or mating behaviour. Limb bones meet the 'axial skeleton' (the axis of the body is the backbone and skull) through connected series of bones called 'girdles' at the shoulder ('shoulder or pectoral girdle') and the hip ('pelvic girdle'). The shoulder and hip girdle bones are broad for attachment of limb muscles. The backbone consists of several vertebrae, which differ in shape according to their function: neck (cervical) vertebrae support the head; trunk (dorsal, or thoracic) vertebrae support the rib cage and have sheets of body muscles attached; lower back (sacral and lumbar) vertebrae support the top of the body and meet with the hip bones; tail (caudal) vertebrae make up the tail, where one is present. Some dinosaurs, such as dromaeosaurs and some ornithopods, even had ossified ligaments supporting their stiff tails.

EVOLUTION OF VERTEBRATES

The origins of dinosaurs within the reptiles is one story, but to understand the broad picture we should delve further back in time to the origin of the first four-legged animals (tetrapods). These early amphibians most likely evolved from lobe-finned fishes called Osteolepiformes in the later half of the Devonian period (more than 370 million years ago). The limb bones of osteolepiform fishes have an identical pattern to those of higher vertebrate animals, with the pectoral fin (arm) having a humerus, ulna and radius, as is found in our own human arms. The first tetrapods were very fish-like, with a similar skull and cheek-bone pattern to that of their ancestral fish-like forms, but differed principally in the presence of digits on the hands and feet. The earliest well-known amphibians—from the Upper Devonian—include an enigmatic form called *Elginerpeton*, known only by some tantalising fragmentary remains from Scotland. *Elginerpeton* is slightly older in age than the better-known *Acanthostega* and *Ichthyostega* from east Greenland. *Acanthostega* is believed to have had a fully aquatic lifestyle, but all early amphibians, even those which invaded the terrestrial habitats, would have still required water for their young to develop, in the same way that modern frogs develop first as tadpoles.

▽ Skull of the Late Devonian amphibian *Ichthyostega*, one of the world's earliest tetrapods

JOHN A. LONG

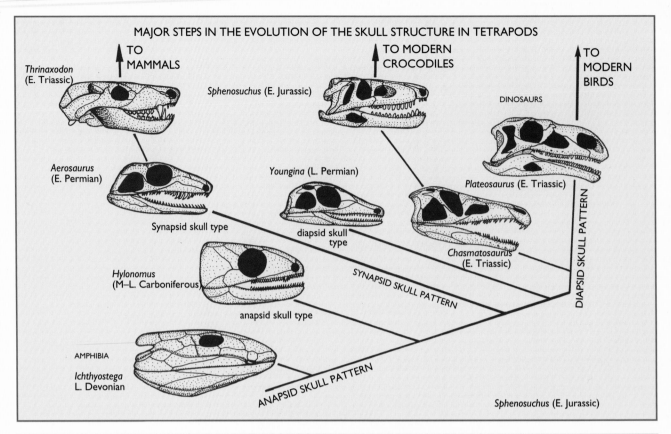

MAJOR STEPS IN THE EVOLUTION OF THE SKULL STRUCTURE IN TETRAPODS

TO MAMMALS

TO MODERN CROCODILES

TO MODERN BIRDS

Thrinaxodon (E. Triassic)

Sphenosuchus (E. Jurassic)

DINOSAURS

Aerosaurus (E. Permian)

Youngina (L. Permian)

Plateosaurus (E. Triassic)

Synapsid skull type

diapsid skull type

Chasmatosaurus (E. Triassic)

DIAPSID SKULL PATTERN

Hylonomus (M–L. Carboniferous)

SYNAPSID SKULL PATTERN

anapsid skull type

AMPHIBIA

Ichthyostega L. Devonian

ANAPSID SKULL PATTERN

Sphenosuchus (E. Jurassic)

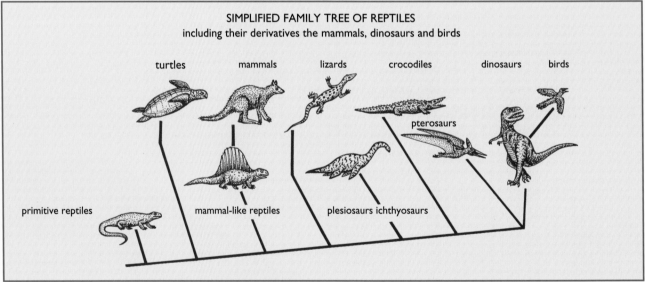

SIMPLIFIED FAMILY TREE OF REPTILES
including their derivatives the mammals, dinosaurs and birds

turtles   mammals   lizards   crocodiles   dinosaurs   birds

pterosaurs

primitive reptiles

mammal-like reptiles

plesiosaurs ichthyosaurs

Australia has yielded a jaw of a Devonian amphibian from near Parkes, New South Wales; it is named *Metaxygnathus*, and is about the same age as the east Greenland amphibians (Campbell and Bell 1977). These primitive amphibians are called labyrinthodonts because of the complex labyrinthine style of infolding of the dentine and enamel in their teeth. It has recently been found that the hand of *Acanthostega* had eight digits, and other Devonian forms had either seven or six digits on the limbs, indicating that the pattern of five fingers per hand and five toes per foot was established in later lineages of amphibians and carried on to all subsequent descendants: reptiles, mammals and dinosaurs. More advanced lineages in some of these groups may then have modified this pattern by having fewer digits, or even losing limbs entirely, as is seen in snakes.

Australia has a highly diverse and well-preserved fauna of Mesozoic labyrinthodonts,

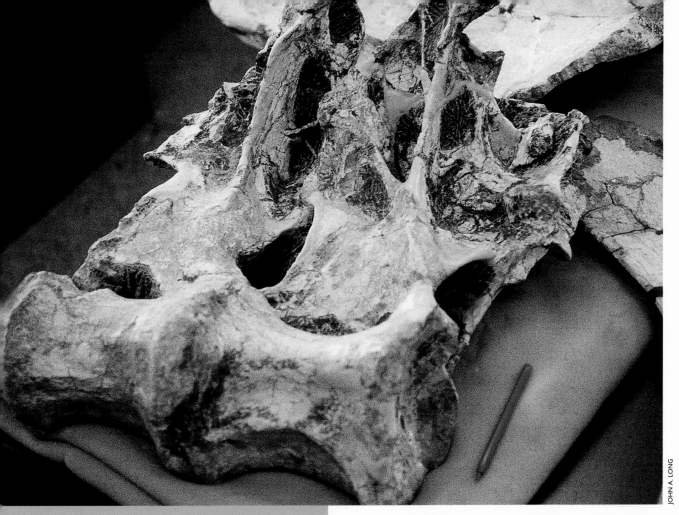

JOHN A. LONG

▲ Vertebrae from a sauropod, showing high neural arches with many complex articulations and cavities (pleurocoels) in the sides of the bones to reduce weight

▼ Linnean levels of classification (left), unnamed levels of classification based on cladistic analysis of groups (right)

KINGDOM	**Kingdom**
Superphylum	Animalia
PHYLUM	**Phylum**
Subphylum	Chordata
CLASS	**Class**
Subclass	Reptilia
Infraclass	taxon: Dinosauria
Superorder	taxon: Saurischia
ORDER	**Order: Theropoda**
Suborder	taxon: Tetanurae
Superfamily	taxon: Coelurosauria
FAMILY	taxon: Maniraptora
SUBFAMILY	**Family: Tyrannosauridae**
GENUS	**Genus:** *Tyrannosaurus*
Subgenus	**Species**: *T. Rex*
SPECIES	
Subspecies	
Tribe, Race	

particularly in the Triassic Period, in which most of the known families are represented. The labyrinthodonts are divided into two major extinct groups, the temnospondyls and the anthracosaurs, each characterised by the structure of their backbones (vertebrae): as in all early amphibians, each segment of the backbone is made up of three parts—an intercentrum at the front, a pleurocentrum behind it, and a neural arch on top. Temnospondyls have large, crescent-shaped intercentra and small, paired pleurocentra. Anthracosaurs have large pleurocentra forming the major part of the backbone (Carroll 1987). There is also a difference in how the skull is attached: in temnospondyls the skull is rigidly fixed to the cheek region, but in anthracosaurs it is loosely attached. All of the Australian Mesozoic amphibians belong to the temnospondyls. The shapes and contact relations of the many bones making up the skulls of these fossil amphibians are very important features used in diagnosing each form.

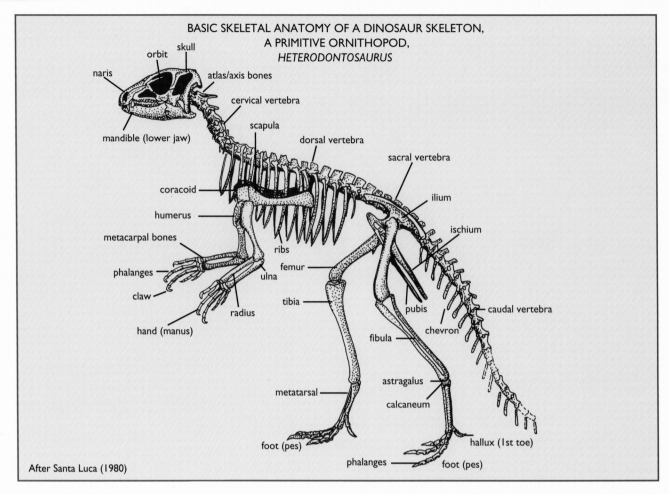

BASIC SKELETAL ANATOMY OF A DINOSAUR SKELETON,
A PRIMITIVE ORNITHOPOD,
HETERODONTOSAURUS

After Santa Luca (1980)

The first reptiles evolved from amphibians approximately 245 million years ago—a conclusion based on the recent find from the Lower Carboniferous rocks near Edinburgh, Scotland, of the skeletons of a small lizard-like beast called *Westlothiana*. It was at that stage that tetrapods may have become independent of the water for the first time. Modern reptiles are distinguished from amphibians by the ability to lay a hard-shelled amniote egg—the young emerging from that egg more or less as a miniature replica of the adult—as well as by several anatomical features of the skeleton. Reptiles and their evolutionary derivatives (dinosaurs, mammals and birds) are, collectively, called the Amniota.

During the latter half of the Palaeozoic Era reptiles diversified into several lineages. The most primitive known reptiles are called anapsids; they have skulls without a hole behind the eyes. Modern turtles and tortoises are living examples of this primitive reptile group. The synapsids, on the other hand, have a single opening (termed the 'temporal fenestra') in the skull behind the eye, low down on the skull wall. Early predators with sails on their backs, such as *Dimetrodon*, are examples of this group. Synapsids reached a peak of diversity in the Late Permian, and it is this group that gave rise to the first mammals, which had appeared by the Middle Triassic (230 million

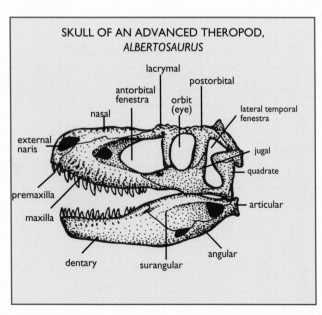

SKULL OF AN ADVANCED THEROPOD,
ALBERTOSAURUS

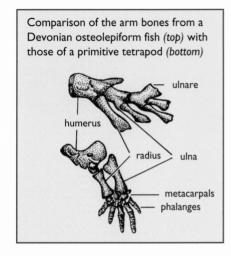

Comparison of the arm bones from a Devonian osteolepiform fish *(top)* with those of a primitive tetrapod *(bottom)*

ulnare
humerus
radius
ulna
metacarpals
phalanges

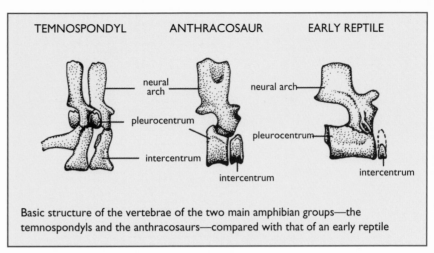

TEMNOSPONDYL ANTHRACOSAUR EARLY REPTILE

neural arch
pleurocentrum
intercentrum

neural arch
pleurocentrum
intercentrum
intercentrum

Basic structure of the vertebrae of the two main amphibian groups—the temnospondyls and the anthracosaurs—compared with that of an early reptile

years ago). In diapsid reptiles the skull has two openings behind the eye. The two major groups of diapsids are the lepidosaurs (lizards, snakes, and some ancient forms such as ichthyosaurs) and the archosauromorphs (crocodilians, pterosaurs, and dinosaurs). Euryapsid reptiles, which include the sea-going plesiosaurians, have a skull with one opening behind the eye high on the skull wall.

DINOSAUR ORIGINS

Dinosaurs are thought to have evolved from thecodonts (swift, running reptiles with teeth set in sockets) at about the same time as the first mammals appeared in the Middle Triassic, after the appearance of the first representatives of our living groups of reptiles (lizards, crocodilians, tortoises, and turtles). Therefore, lizards and tortoises are regarded as primitive reptilian groups which have survived on to the present day, with snakes evolving from lizards later (about 100 million years ago).

Thecodonts share with dinosaurs the feature of an opening in the skull in front of the eye (termed 'antorbital fenestra'). Thecodonts were the dominant reptile group throughout the Triassic Period, and it may have been their ability to run for short distances on their hind legs that gave rise to dinosaurs. This particular running ability used a strengthened lower leg joint in which compact, strong metatarsal bones transferred most of the weight of the animal down from the ankle to the toes. Birds probably evolved from small carnivorous dinosaurs (coelurosaurs or ancestral dromaeosaurs) some time before the Middle Jurassic. Dinosaurs and mammals should be equated as vertebrate groups distinct from the reptiles we know today. Birds differ from these in being a specialised subgroup within the dinosaurs that developed the ability to fly.

Dinosaurs arose just before the Middle Triassic (some 230 million years ago), and by the end of the Triassic several of the major groups of dinosaurs had become established all around the

▼ The Thorny Devil, *Moloch horridus*, a living primitive lizard from central Australia. Lizards have a sprawling gait whereas dinosaurs walked with their limbs vertically oriented.

world. Dinosaurs belong to the Archosauro-morpha, a group also including crocodilians, pterosaurs, and some extinct reptilian groups, whose members share the basic specialised type of ankle joint that facilitates a more upright walking posture. Thus, the astragalus and calcaneum bones of the foot have broad articulation surfaces for their contact with each other, and the other foot bones are also modified towards assisting an upright gait. Primitive reptile groups that evolved well before dinosaurs generally retained a sprawling type of gait with the arms and legs sticking sideways out from the body.

Dinosaurs are further distinguished from other archosauromorph groups by having joints in the limbs for a nearly vertical limb alignment when viewed from the front. Dinosaurs have a 'mesotarsal' ankle joint where the line of flexion passes between the proximal and distal tarsal bones. The metatarsal bones thus carry the weight of the animal vertically to the feet, acting as an extra jointed limb in the legs. The foot digits form the main surface in contact with the ground—a condition, not surprisingly, also seen in birds. Thus, probably right from the start, dinosaurs had the inbuilt capability for walking on two hind legs (bipedalism), and several dinosaur lineages evolved to take advantage of this style of locomotion. It is suggested by some scientists that the dinosaurs which walked on four limbs (quadrupedal forms) may have evolved from earlier dinosaurs that had walked upright and then later reverted to a quadrupedal gait.

CLASSIFICATION OF DINOSAURS

The modern approach to looking at the evolutionary relationships of organisms and forming appropriate classifications for them is called 'cladistics'. In this methodology the character states found in different species are analysed to determine whether they are 'primitive' or 'advanced' within that group. An example of a primitive character is one that might occur at the beginning of their evolutionary radiation and is widespread in all members of the group (for example, a feature common in all dinosaurs is walking with the hind limbs vertical and on the metatarsal bones). Advanced characters (also labelled 'derived') are limited to smaller specialised groups which evolved them for a specific adaptation, and thus unite them as having a common evolutionary origin. Examples of specialised character states are the long neck and gigantic size of large sauropod dinosaurs, or features like the enlarged sickle-claw on the foot of dromaeosaurids. Diagrams showing the schemes of relationships based on the analysis of their specialised character distributions are called 'cladograms'. Some of these, showing the latest hypotheses of dinosaur interrelationships, are shown in this book. However, to exploit the scientific reasoning behind how such schemes are made is beyond the scope of this book as it requires very specialised and detailed knowledge of the anatomy of dinosaurs.

ANKLE-JOINT EVOLUTION IN REPTILES
leading to the first dinosaurs

Galesphyrus,
primitive reptile

Euparkeria,
thecodont

Lagosuchus,
theocodont

Struthiomimus,
dinosaur

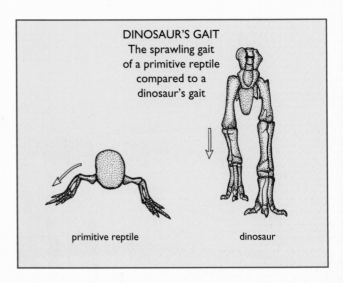

DINOSAUR'S GAIT
The sprawling gait of a primitive reptile compared to a dinosaur's gait

primitive reptile   dinosaur

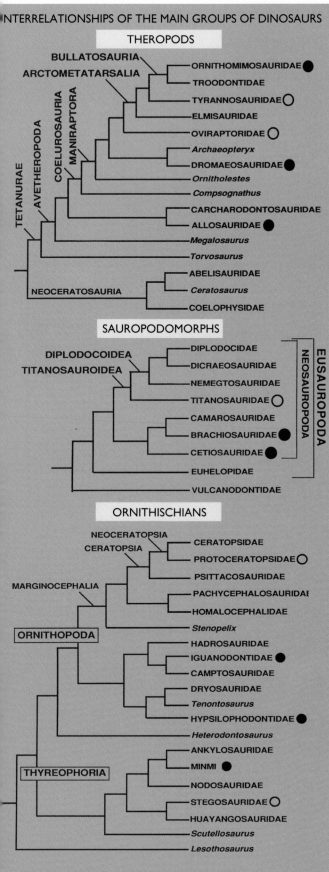

INTERRELATIONSHIPS OF THE MAIN GROUPS OF DINOSAURS

THEROPODS

BULLATOSAURIA
ARCTOMETATARSALIA

ORNITHOMIMOSAURIDAE ●
TROODONTIDAE
TYRANNOSAURIDAE ○
ELMISAURIDAE
OVIRAPTORIDAE ○
Archaeopteryx
DROMAEOSAURIDAE ●
Ornitholestes
Compsognathus
CARCHARODONTOSAURIDAE
ALLOSAURIDAE ●
Megalosaurus
Torvosaurus
ABELISAURIDAE
Ceratosaurus
COELOPHYSIDAE

TETANURAE
AVETHEROPODA
COELUROSAURIA
MANIRAPTORA

NEOCERATOSAURIA

SAUROPODOMORPHS

DIPLODOCOIDEA
TITANOSAUROIDEA

DIPLODOCIDAE
DICRAEOSAURIDAE
NEMEGTOSAURIDAE
TITANOSAURIDAE ○
CAMAROSAURIDAE
BRACHIOSAURIDAE ●
CETIOSAURIDAE ●
EUHELOPIDAE
VULCANODONTIDAE

NEOSAUROPODA
EUSAUROPODA

ORNITHISCHIANS

NEOCERATOPSIA
CERATOPSIA

CERATOPSIDAE
PROTOCERATOPSIDAE ○
PSITTACOSAURIDAE
PACHYCEPHALOSAURIDAE
HOMALOCEPHALIDAE
Stenopelix

MARGINOCEPHALIA

ORNITHOPODA

HADROSAURIDAE
IGUANODONTIDAE ●
CAMPTOSAURIDAE
DRYOSAURIDAE
Tenontosaurus
HYPSILOPHODONTIDAE ●
Heterodontosaurus

THYREOPHORIA

ANKYLOSAURIDAE
MINMI ●
NODOSAURIDAE
STEGOSAURIDAE ○
HUAYANGOSAURIDAE
Scutellosaurus
Lesothosaurus

Note: The groups represented in Australia are marked by a black circle, and those possibly here (either by footprints or dubious specimens) by an open circle.

Dinosaurs are classified according to the structure of the pelvis: the two main groups are the saurischians (Order Saurischia), or lizard-hipped dinosaurs, and the ornithischians (Order Ornithischia), or bird-hipped dinosaurs. Saurischians have the primitive pattern of pelvic bones with a forward-projecting pubis and a rearwards-directed ischium. This is the type of pelvis seen in reptiles that arose before dinosaurs, and can be seen in the surviving groups today, such as lizards and crocodiles. Ornithischians are specialised in that the ischium and pubis point backwards and are close together. Ornithischians are further specialised in that the lower jaw has a unique bone, the predentary, situated where the two halves of the lower jaws meet. The predentary is toothless but may have been covered by a horny beak like that which turtles have.

The saurischian dinosaurs share many primitive archosaurian features and it is not known whether they share a common ancestry from one lineage. There are two main groups of Saurischia: the Theropoda, or flesh-eating, bipedal dinosaurs; the Sauropodomorpha, which includes the giant, long-necked plant-eaters and related ancestral forms, such as prosauropods (for example, *Plateosaurus*, *Agrosaurus*). Theropods include several higher groups, among them neoceratosaurians *(Coelophysis, Dilophosaurus, Ceratosaurus, abelisaurids), allosaurids and carcharodontosaurids (Allosaurus, Giganotosaurus,*

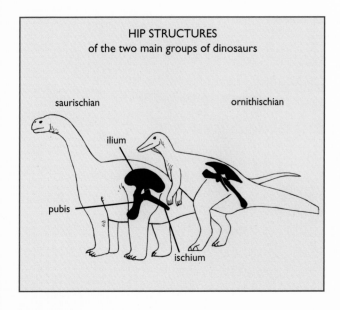

HIP STRUCTURES
of the two main groups of dinosaurs

saurischian

ornithischian

ilium

pubis

ischium

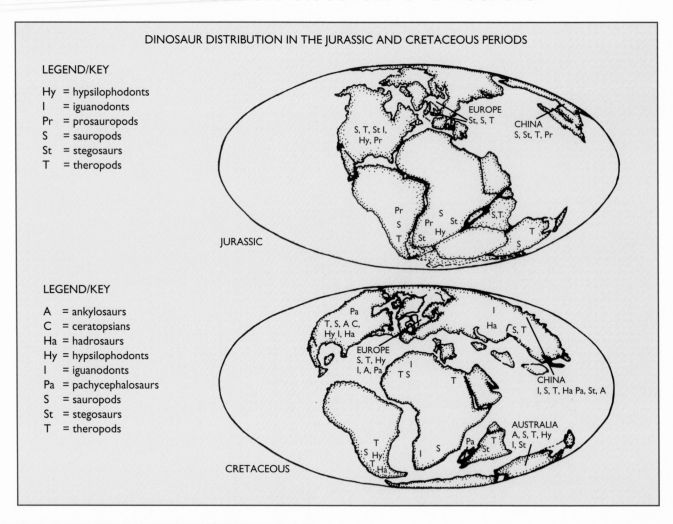

DINOSAUR DISTRIBUTION IN THE JURASSIC AND CRETACEOUS PERIODS

LEGEND/KEY

Hy = hypsilophodonts
I   = iguanodonts
Pr = prosauropods
S   = sauropods
St = stegosaurs
T   = theropods

EUROPE
St, S, T

CHINA
S, St, T, Pr

S, T, St I,
Hy, Pr

Pr
S
Pr
S
St
Hy
St
S
St
S,T.
T
S

JURASSIC

LEGEND/KEY

A   = ankylosaurs
C   = ceratopsians
Ha = hadrosaurs
Hy = hypsilophodonts
I   = iguanodonts
Pa = pachycephalosaurs
S   = sauropods
St = stegosaurs
T   = theropods

Pa
T, S, A C,
Hy I, Ha

EUROPE
S, T, Hy
I, A, Pa

I

Ha
S, T

I
T S

T

CHINA
I, S, T, Ha Pa, St, A

T
S
Hy.
T.
Ha

I
S
Pa
T
St

AUSTRALIA
A, S, T, Hy
I, St

CRETACEOUS

▼ Major Mitchell's cockatoo, *Cacatua leadbeater*. It is thought the ancestors of birds evolved from small carnivorous dinosaurs some time before the Middle Jurassic

KEN GRIFFITHS

Carcharodontosaurus, Acrocanthosaurus), coelurosaurs and maniraptorans (dromaeosaurids, *Archaeopteryx,* oviraptorids, tyrannosaurids, ornithomimosaurids, and others). There are several groups of primitive theropods which have hands with four fingers, and subsequent advanced groups such as the allosaurids (with three fingers per hand) and the tyrannosaurids (with only two digits per hand). Bizarre forms such as *Mononykus* have only one digit per hand, and are now considered as primitive birds, or highly derived forms of theropod dinosaurs. Theropods are specialised also in the ankle joint with an astragalus bone having a high ascending process in contact with the shin bone (tibia), and stout, closely compacted tarsal bones of the foot in advanced forms (the 'arctometatarsalian' condition, as is found in tyrannosaurids).

The Sauropodomorpha includes two major groups: the prosauropods (plateosaurids, melanorosaurids, massospondylids, thecodontosaurids, anchisaurids, blinkanosaurids) and the well-known, gargantuan, long-necked Sauropoda. Prosauropods were abundant in the late-Triassic and Jurassic periods. Sauropods were the largest land animals to have ever lived, some reaching gargantuan sizes, like *Brachiosaurus,* with estimated weights of 80 tonnes, or the massive *Argentinosaurus* (from South America), estimated to be more than 30 m long and weighing maybe up to 90 tonnes. The main sauropod families are the vulcanodontids, diplodocids, brachiosaurids, cetiosaurids, camarosaurids, nemegtosaurids and titanosaurids. Most of the sauropods were Jurassic animals, with only some groups, particularly the titanosaurids, surviving through to the Late Cretaceous.

The Ornithischia contains several groups, some bipedal and some quadrupedal, but all were planteaters. Small, agile, running dinosaurs like the hypsilophodonts are well-represented all over the world, being the equivalent of the gazelles in the dinosaur kingdom (Galton 1974). Iguanodontids were large, bipedal plant-eaters with spiked thumbs. Hadrosaurs, or duck-billed dinosaurs, were bipedal forms, some with elaborate head processes like helmets, crests and crowns, presumably evolved for the indifferent acoustic properties. The hadrosaurs had duck-like bills and many teeth for grinding down coarse plant material, some having as many as 60 tooth positions per jaw, with 700 teeth exposed at one time. Ceratopsians were quadrupedal forms with an armoured frill covering the neck and often possessing horns for mating rituals or defending themselves against carnivores—*Triceratops* being one of the better-known examples. Pachycephalosaurs were bipedal ornithischians with thick, domed skulls, possibly used for ramming each other during mating battles. Armoured ornithischians, called thyreophorans, include two very well-known groups: the stegosaurs with heavy bony plates or rows of spikes along the back; and the ankylosaurs, or tank-like dinosaurs, with thick, armoured scutes on the back and most of which had a formidable clubbed tail.

Dr Paul Sereno (1986) gives a detailed hypothesis of the evolutionary relationships of ornithischian dinosaurs. Sereno believes that the stegosaurs and ankylosaurs arose from one lineage that was ancestral to most other ornithischians. The ornithopods (including dryosaurs, hypsilophodonts,

▽ The Cretaceous–Tertiary boundary exposed at Waimakariri Gorge, Canterbury, New Zealand. Preservation of this layer in different parts of the world enables us to reconstruct the final days of the dinosaurs

DR EWAN FORDYCE

iguanodonts and hadrosaurs) arose next, with pachycephalosaurs and ceratopsians representing the most advanced ornithischian lineages (united as the 'marginocephalians', for they share a frill-like margin on the head).

DISTRIBUTION OF DINOSAURS WORLDWIDE

Only a few of the many known families of dinosaurs have been recorded from Australia and New Zealand. It is interesting to see which groups are absent and to try and determine whether these absences are real, occurring because the groups did not inhabit the Australasian region, or whether they are artificial, because so few dinosaurs have actually been found in the Australasian region and some absent groups may yet be found.

The saurischians are well-represented all over the world, and in the Australasian region there are theropods and sauropods, as well as a possible prosauropod, although its exact locality is dubious. These groups probably arose sometime late in the Triassic and may have migrated through Pangaea to reach all corners of the supercontinent before its subsequent break-up. However, certain families of theropods have restricted distributions, such as troodontids and abelisaurids. Recent discoveries of possible oviraptorosaurs and ornithomimosaurids in Australia, earlier in time than for Asia and North America, suggest that these groups may have had an origin in Gondwana. However, further indisputable remains need to be found before such hypotheses can be confirmed.

The ornithischians include several groups with restricted distributions because some of these evolved after the initial separation of Gondwana from Pangaea, which isolated the groups and prevented them from reaching some of the newly formed continents. The discovery of a possible neoceratopsian arm bone from Victoria lends weight to the idea that this group may have appeared first in Gondwana, before migrating north to Asia and North America, where the

majority of the group thrived in Late Cretaceous times. The stegosaurs are known from Europe, America, Africa and China in the Jurassic, but only in India by the Cretaceous, as the group became isolated when India split off from Gondwana at the end of the Jurassic.

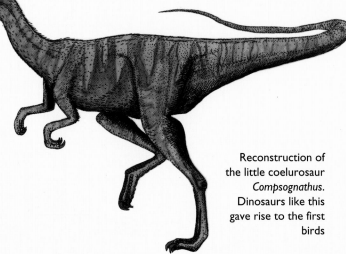

Reconstruction of the little coelurosaur *Compsognathus*. Dinosaurs like this gave rise to the first birds

The factors affecting the decline of the stegosaurs elsewhere—perhaps the rise of a specific type of predator, or a change of vegetation or climate—were presumably absent from the Indian subcontinent as it drifted northwards from Gondwana at the end of the Jurassic. The existence of stegosaurs in India, which adjoined Australia in the Jurassic, suggests strongly that the group could have invaded Australia. Footprints of a possible stegosaur have recently been identified from near Broome, Western Australia (see chapter 6), indicating that the group may have made it into eastern Gondwana before India broke away. The other ornithischians which achieved widespread distribution were the ankylosaurs, iguanodontids and hypsilophodontids, all of which inhabited Asia, Africa, Europe, North and South America and the Australasian region.

EVOLUTION OF BIRDS FROM DINOSAURS

Knowledge of the first bird, *Archaeopteryx*, is based on several well-preserved skeletons, some with feather impressions clearly preserved, found in the Late Jurassic Solnhofen Limestone of Germany. It was at first distinguished from similarly sized small coelurosaurian dinosaurs only by the presence of feathers and the relatively longer forearms. Since the discovery in 1996 of the first feathered dinosaur, *Sinosauropteryx prima*, from the Early Cretaceous of China (Gibbons 1996), and the discovery of *Mononykus*, a bird resembling a theropod dinosaur, from the Late Cretaceous of Mongolia, the evolutionary gap between dinosaurs and birds has completely disappeared (Chiappe *et al.* 1996). In all other anatomical features *Archaeopteryx* is simply a small theropod dinosaur, and all birds are now thought to have evolved from an *Archaeopteryx*-like ancestor. Dr Tony Thulborn (1984) argued a case for *Archaeopteryx* as being less of a bird than the giant meat-eating *Tyrannosaurus rex* is, since *T. rex* shows all the skeletal features of *Archaeopteryx* shared with birds, as well as additional bird-like features not seen in *Archaeopteryx*; not all palaeontologists agree with this hypothesis, however. *Archaeopteryx* nowadays becomes just another small, feathered dinosaur, with the first 'true birds' being those capable of proper avian flight, and possessing the necessary skeletal specialisations for flight not found in any dinosaurs. The first true flying birds, in this sense, are possibly the enantiornithines, one of which *(Nanantius),* comes from the Early Cretaceous of Queensland (see chapter 7).

The evolution of feathers from reptilian scales is not seen nowadays as much of an anatomical hurdle in the story of bird origins; for, simply by administration of retinoic acid (Vitamin A) at a particular stage of their embryonic development, chick embryos can develop feathers where scales normally grow (on the legs), and develop scales over the body (Thulborn 1985a). Thus, the reptilian ancestor of birds had the inbuilt potential to develop feathers. One can only suppose that it could have been changes in developmental rate or diet, in conjunction with an adaptational need to regulate body temperature, that triggered off the need for the development of feathers. The ability to fly may have come next as feathered running dinosaurs used their now broader forelimbs to direct swarms of insects up past the mouth. These swift runners may have scampered up trees to escape predators, and then utilised the feathered body surfaces to slow down their fall when jumping from heights. This activity could quickly select out individuals with a larger surface area of feathers and with longer forearms, leading to a preflight condition. Many groups of non-flying creatures develop an ability to glide in this way, such as certain frogs, lizards, and gliding possums.

The numerous skeletal similarities between birds and reptiles, particularly coelurosaurian dinosaurs, clearly indicates that these groups came from a

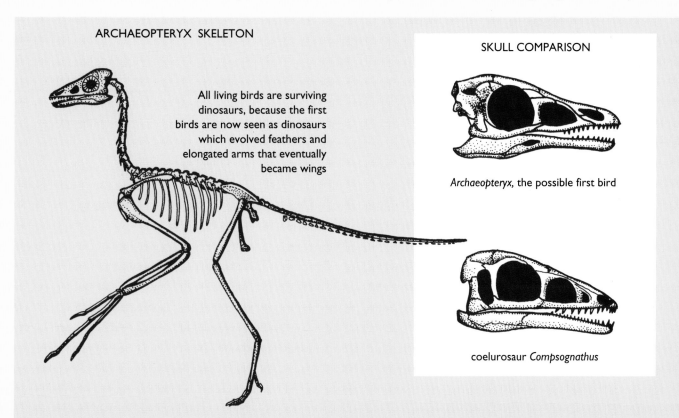

ARCHAEOPTERYX SKELETON

All living birds are surviving dinosaurs, because the first birds are now seen as dinosaurs which evolved feathers and elongated arms that eventually became wings

SKULL COMPARISON

Archaeopteryx, the possible first bird

coelurosaur *Compsognathus*

▲ Dinosaur extinctions at the end of the Cretaceous Period may have been assisted by the consequences of a giant asteroid impact, which would have caused global fires, earthquakes, volcanic eruptions, acid rain and prolonged darkness and created longer-term effects such as global warming (greenhouse effect)

▶ The Transantarctic Mountains, Antarctica, have yielded dinosaur remains from the time when Antarctica, with Australasia, formed part of one supercontinent, Gondwana

JOHN A. LONG

common ancestor. The remarkable similarities between *Archaeopteryx* without its feathers and some of these dinosaurs—*Compsognathus*, for instance—makes it difficult to distinguish them if only part of the skeleton is found. Most palaeontologists now consider all living birds 'living dinosaurs' as they are all now regarded as descendants of early theropod dinosaurs.

EXTINCTION OF DINOSAURS: CURRENT THEORIES

At the end of the Cretaceous Period (66 million years ago) the dinosaurs along with some other groups of animals died out. However, this event was not as sudden as some books make out; instead, the dinosaurs had been declining in both species and numbers for some few million years beforehand. On land the dinosaurs, both large and small, died out but other reptiles (such as the lizards, snakes, crocodilians and tortoises), together with small mammals and birds, have continued on to the present day. Similarly, in the seas the great marine reptiles—the ichthyosaurs, plesiosaurs and mosasaurs—died out, but turtles and most of the fish groups survived. It is clear now that ichthyosaurs actually died out well before the end of the Cretaceous.

The extinctions affected all groups of animals

right down to microscopic plankton: in nearly all cases there are examples of families which did not survive the end of the Cretaceous Period, and of those that did. What brought about such selective extinctions? Why did the small dinosaurs also die out but not the mammals or the birds? These are questions that have puzzled scientists for many years and, unfortunately, still continue to do so despite many suggested solutions that have been put forward.

It is interesting to note that as far back as the 1950s some scientists recognised the potential of massive asteroid impacts on the Earth as having the destructive capability to cause extinction events. M. W. de Laubenfels (1956) states: 'Attention is called to the great destruction that resulted from a meteorite impact in Siberia in 1908. A larger impact would cause more widespread destruction. Several larger impacts may have occurred in geologic time. The survivals and extinctions at the close of the Cretaceous are such as might be expected to result from intensely hot winds such as would be generated by extra large meteoric or planetesimal impacts. It is suggested that, when the various hypotheses as to dinosaur extinction are being considered, this one be added to the others.'

In recent years this theory has gained much support from physicists, but not from palaeo-

4 3

ntologists. It is based largely on the evidence of a high level of the platinum group element iridium—an element occurring in meteorites at levels many times greater than the average levels normally found in the Earth's crust—present throughout the world in sediments at the end of the Cretaceous Period. In the same layer with the high iridium levels, large numbers of grains of shocked quartz crystals occur, believed to be a byproduct of the massive impact. In addition, the presence of high levels of carbon in the sediments just above this layer implies that huge fires raged at the time, worldwide. The position of the impact event has been identified as the Chicxulub crater, in the Yucatan district of Mexico. Hildebrand (1997) argues the effect on the environment would have been devastating from such an impact: it would cause acid rain from shocked atmospheric nitrogen or released sulphur dioxide, followed by cooling due to stratospheric sulphate aerosols, and then greenhouse-effect warming due to released carbon dioxide. A global dust cloud would also have been created, which may have lasted more than three months, causing darkness and thus disrupting photosynthesis in plant communities. Overall, this would have put the whole ecosystem into chaos, with strong selection pressures on the organisms to cope with the

sudden dramatic changes in climate, atmosphere, and vegetation. Nevertheless, this does not adequately explain why some organisms survived, both cold-blooded and warm-blooded creatures of all sizes, while others did not.

The biological approach to the extinctions mystery favours a less dramatic scenario. In this view, the long-term climatic changes caused by regular fluctuations in the Earth's orbit, and changes brought about by continental movement, had resulted in a gradual change in the composition of the ecosystem; in the course of this dinosaurs were gradually being displaced by smaller mammals or birds, which probably adapted more readily to feeding at night, or even preying on dinosaur eggs. Changes in vegetation at this time saw the rise of flowering plants (angiosperms), which partly displaced the gymnosperms (for example, conifers, cycads); this may have affected the plant-eating dinosaurs if new poisons or substances had developed in these new plants. Many other biological arguments have been put forward to explain the extinctions, such as overkill by too many predators, the senility of certain races of animals, increasing genetic malformations, the co-evolution of parasites and diseases, and even sterilisation of dinosaurs by excessive heat or toxins. However, none of these fully explains the selection of certain groups of animals and plants across the Cretaceous–Tertiary boundary.

In summary we do not have any clear explanation for the extinctions; it seems, too, that the more research that is done on the subject, the more questions there are to answer (Molnar and O'Reagan 1989). My personal viewpoint is that near the end of the Cretaceous there were many drastic changes in the environment: a decline in oxygen levels had taken place since the Early Cretaceous; there was the rise of flowering plants replacing the earlier gymnosperm-dominated flora; there were large changes in sea-levels, and right at the end there was a large asteroid impact. Such an onslaught on the environment from so many different sources meant something had to give; in this case, it was the dinosaurs.

three

AUSTRALIAN AND NEW ZEALAND DINOSAUR DISCOVERIES: A BRIEF HISTORY

FIRST FINDS IN AUSTRALIA

The first dinosaur bones found in Australia purportedly came from the northeastern coast, somewhere in Queensland, and were collected by crew members from the HMS *Fly* in 1844. The specimens consisted of a few associated bones from a primitive dinosaur. These were purchased by the British Museum (Natural History) in 1879, but not studied until 1891, when Seely described the bones as belonging to a new type he named *Agrosaurus macgillivrayi*. Unfortunately, the ship's log has no record of what locality the fossils came from, and to date there have been no other dinosaur bones collected from the northeastern coast of Queensland; so, it is difficult to be certain about *Agrosaurus* even being an Australian dinosaur. The discovery of fossil marine reptiles like the plesiosaurs and ichthyosaurs predates any large dinosaur find. The first discovery of ichthyosaurs in Australia was made by James Sutherland in 1865, on the Flinders River in north central Queensland. Sutherland's specimens of ichthyosaur vertebrae were sent to Professor Frederick McCoy, at the Museum of Victoria, who described them in a short note in 1867.

On his return to the site, Sutherland uncovered a more complete specimen, including a skull and many more vertebrae, all of which was also sent to Melbourne for McCoy to study. Despite the fact that McCoy wrote a second paper describing the new material (1869), it was not detected that two skulls were actually present in the material (Wade 1984).

William Hann led an expedition to the north of Queensland in 1872, with geologist Norman Taylor. They found bones of an ichthyosaur in the Walsh River area. They collected quite a few of the bones, which made their way back to the Queensland Museum, although others were buried beneath the ashes of their campfire. Other Queensland ichthyosaur fossils were found at Marathon Station and described by Etheridge (1888) as 'Ichthyosaurus marathonensis'. In 1892 Jack and Etheridge compiled a large book on the geology and fossils of this region, in which they mentioned finds of fossil reptiles from Queensland. In 1897 Etheridge described a plesiosaur preserved in opal from White Cliffs, New South Wales, and in 1904 described further remains from the same site. These specimens were purchased from opal dealers.

The first discovery of a dinosaur from a well-documented Australian site is that of a small claw from a carnivorous dinosaur found on the coast of eastern Victoria, near Cape Paterson. The bone, affectionately dubbed 'the Cape Paterson claw' was described in 1906 by Sir Arthur Smith Woodward, of the British Museum (Natural History), as belonging to a close relative of the English *Megalosaurus*. This site was to remain barren for many decades until the late 1970s, when interest in the area picked up as further dinosaur discoveries were made (which are described below).

In 1899 a scrap of large reptile jawbone was given to the Queensland Museum by Mr A. Crombie of Hughenden. This was later described (in 1924 and 1930) as the giant plesiosaur *Kronosaurus* by Mr Heber Longman, who was a self-trained palaeontologist at the Queensland Museum and later became its director. In 1924

Mr Thomas Jack and Mr Wood informed Longman that a large fossil reptile skeleton was exposed on Durham Downs Station in the Roma district. Fragments of bone were forwarded to the museum by the station manager, Mr A. Browne, and soon afterwards the bones were confirmed to be from a new type of dinosaur. Browne further assisted the Queensland Museum by taking Longman to the site and, together, they collected much more material. In 1926 and in 1927 Longman published two papers describing the bones as being from a new type of large sauropod dinosaur, which he named *Rhoetosaurus brownei*. However, it was not until 1975 that the Queensland Museum revisited the site and Dr Mary Wade uncovered several more bones, including a foot of *Rhoetosaurus*, belonging to the original specimen.

Another large sauropod dinosaur, *Austrosaurus*, was described in 1933 by Longman. The remains of this creature were first noticed by a Mr H. Wade, and drawn to the attention of Mr H. McKillop, manager of Clutha Station, near Maxwelton in northern Queensland. McKillop later showed the remains to his brother, Dr M. McKillop, who sent the specimens to the Queensland Museum. When these important discoveries were being made, the Queensland Museum was short of funds and could not afford to readily send dinosaur specialists out to such remote locations. Most of the important dinosaur finds were thus made by local people who excavated the bones as best they could and sent them in to the Queensland Museum.

Other museums at that time did have funds to collect fossils in remote locations. In 1923–24 the British Museum (Natural History) sent Sir George Hubert Wilkins to collect fossil reptiles and ammonites in Queensland, some of which were later donated to the Queensland Museum. Perhaps the greatest loss was to occur in the early 1930s, when an American expedition from the Museum of Comparative Zoology, Harvard University, visited Australia in 1931–32. The leader of the expedition, Dr W. E. Schevill, approached the director of the Australian Museum to see if the museum wanted to

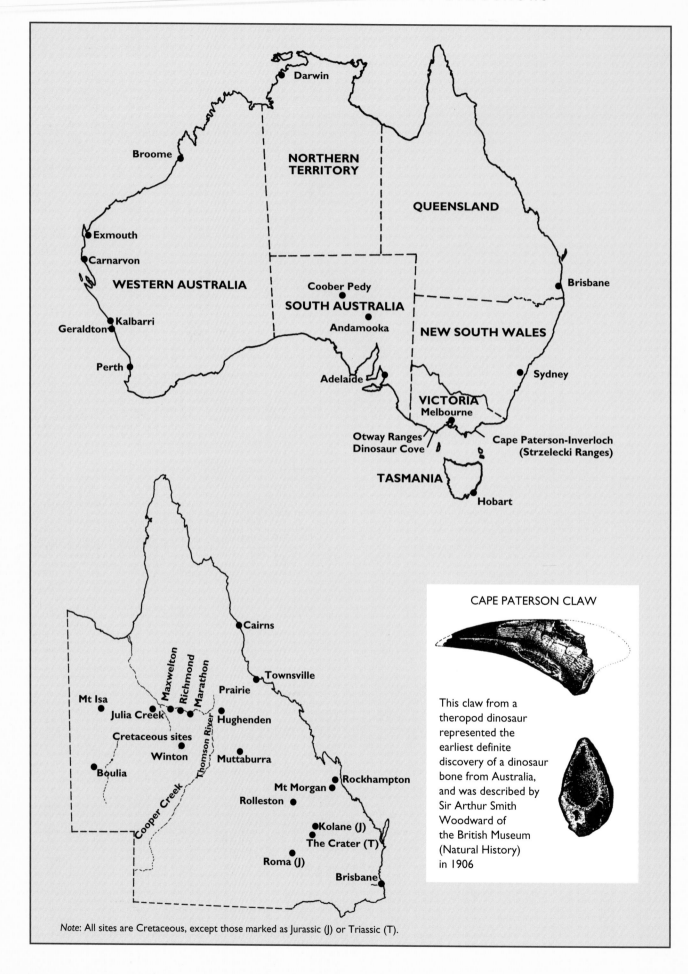

Darwin

NORTHERN
TERRITORY

Broome

QUEENSLAND

Exmouth

Carnarvon

WESTERN AUSTRALIA

Coober Pedy

SOUTH AUSTRALIA

Brisbane

Kalbarri

Geraldton

Andamooka

NEW SOUTH WALES

Perth

Adelaide

Sydney

VICTORIA
Melbourne

Otway Ranges
Dinosaur Cove

Cape Paterson-Inverloch
(Strzelecki Ranges)

TASMANIA

Hobart

Cairns

Maxwelton
Richmond
Marathon

Townsville

Prairie

Mt Isa

Julia Creek

Hughenden

Cretaceous sites

Thomson River

Winton

Muttaburra

Boulia

Rockhampton

Mt Morgan

Cooper Creek

Rolleston

Kolane (J)

The Crater (T)

Roma (J)

Brisbane

CAPE PATERSON CLAW

This claw from a
theropod dinosaur
represented the
earliest definite
discovery of a dinosaur
bone from Australia,
and was described by
Sir Arthur Smith
Woodward of
the British Museum
(Natural History)
in 1906

Note: All sites are Cretaceous, except those marked as Jurassic (J) or Triassic (T).

send a representative palaeontologist along to collect material, but no interest was shown in participating in the expedition. On that trip the expedition found a snout of a juvenile *Kronosaurus* in Grampian Valley, 50 km north of Richmond. A local resident told Schevill that a further 8 km to the north, in the Army Downs region, there was a large skeleton. The specimen was found articulated in a series of 15 limestone nodules buried in the ground. Schevill's assistant, known as 'the maniac' because of his fondness for the use of explosives, dynamited the bones out. They were later loaded onto a truck, and eventually, 4 tons of rock were shipped to the United States, each block wrapped in sheep fleece.

The museum director at Harvard was horrified at the specimens being wrapped like this and had each one washed in disinfectant because of the possible presence of anthrax. After many years of preparation by Mr T. E. White, the skeleton was found to be the giant pliosaur *Kronosaurus,* although there is still some doubt as to whether the specimen really is identical to the snout described earlier by Longman (see chapter 7 for a detailed explanation). A reconstruction of the skeleton, with the missing parts restored after another pliosaur, was eventually put on display at Harvard's Museum of Comparative Zoology in 1959, although the skeleton itself has not yet been fully described. Other blocks containing bones collected from Queensland by Schevill's expedition also await further preparation.

This was the end of a glorious era of discovery. It could not have been summed up better than in Heber Longman's own words, at the conclusion of his first paper in 1926 describing Australia's first big dinosaur discovery, that of *Rhoetosaurus*: 'One may forecast, however, that specimens will eventually be discovered in Australia comparable to the almost-complete skeletons found in America and Europe. Probably it will also be found that the bulky, herbivorous *Rhoetosaurus* had to contend with carnivorous dinosaurs of the *Tyrannosaurus* type, and many new forms will be made known from Australian deposits'—and, indeed, the discoveries were made, but not for another five decades or so.

◁ Heber Albert Longman, palaeontologist and director of the Queensland Museum during the dinosaur discovery days in the first half of the twentieth century. Despite having no formal training in science, Longman went on to describe some of Australia's most spectacular dinosaurs and Mesozoic marine reptiles

◁ Tom Marshall, of the Queensland Museum, with the reconstructed leg bone of *Rhoetosaurus brownei,* sent in to the Queensland Museum in 1924

◁ The site where *Rhoetosaurus* was found— nowadays on Taloona Station, near Roma, Queensland

◁ Dr William Schevill, from Harvard University, who led the expedition to Queensland that discovered the near-complete skeleton of *Kronosaurus* in 1932

MODERN DISCOVERIES IN QUEENSLAND

Since the 1970s palaeontologists at the Queensland Museum and their colleagues at the University of Queensland have been actively collecting from the earlier known sites and discovering new dinosaur sites in north central Queensland. Much of the impetus began with Dr Alan Bartholomai, who was appointed director of the Queensland Museum in 1969.

One of the most spectacular of the recent dinosaur discoveries was that of the large ornithopod *Muttaburrasaurus* by Dr Bartholomai and Queensland Museum entomologist, Dr Edward Dahms. The skeleton was brought to the attention of the museum by a grazier, Mr D. Langdon. It was collected in 1963 from many fragments, in a cattle yard on the Thomson River near Muttaburra, central Queensland. Many other pieces had been souvenired by locals and a public appeal soon saw many of these returned to the museum. The lengthy job of preparing the skeleton was sponsored by the cereal company, Kelloggs, and, as a result of this sponsorship there are several copies of the mounted, restored skeleton on public display in museums around Australia today.

Further collecting by Queensland Museum teams resulted in other large sauropod dinosaur material from Hughenden, collected by Bartholomai in 1959 from Alni Station, and described in detail by Coombs and Molnar (1981). In 1964 Bartholomai collected the remains of a small ankylosaur from a location near Roma, and this was named as a new type, *Minmi paravertebra*, by Dr Ralph Molnar (1980a). In recent years there have been several new finds collected from the Mesozoic rocks of Queensland which are being prepared and awaiting detailed study, including a second skull of *Muttaburrasaurus* (collected on Dunluce Station by Dr Mary Wade in late 1987), and an almost complete new skeleton of *Minmi* (discovered from near Richmond in late 1989).

As a result of the discovery of dinosaur footprints about 120 km southwest of Winton by Mr Ron McKenzie, Thulborn and Wade went to the site to make a detailed study. The footprint site had been visited earlier, in 1971, by American palaeontologist Dr R. Telford with Bartholomai, who determined that the layer of rock containing the dinosaur footprints extended to a second site about 100 m away. Thulborn and Wade were assisted by a large number of volunteer helpers in

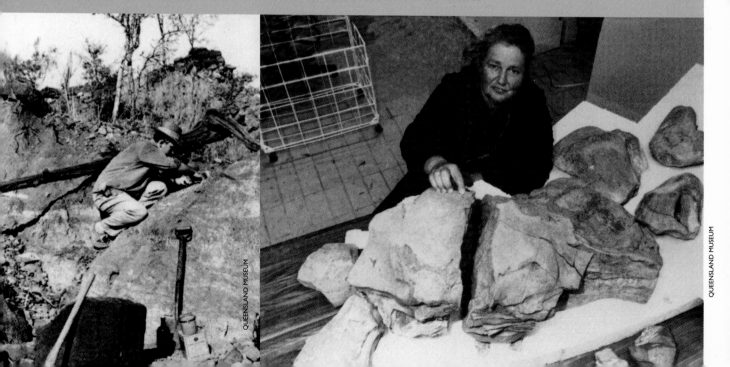

▽ Dr Alan Bartholomai, Director of the Queensland Museum, the man responsible for finding and excavating many Queensland dinosaur fossils, including *Minmi* and *Muttaburrasaurus*

▷ Dr Mary Wade of the Queensland Museum with the Dunluce skull of *Muttaburrasaurus*

▲ Dinosaur tracks from Lark Quarry, near Winton, Queensland. Here over 3300 dinosaur footprints, representing three different species, are exposed on a single bedding surface

◄ Dr Anne Warren and Dr Mark Hutchinson with the skeleton of the Jurassic labyrinthodont *Siderops*

the preparation of the footprint sites, which revealed some 3300 dinosaur footprints, representing over 150 individuals of three different species. This site (Lark Quarry, near Winton) is now famous as a tourist attraction.

Fieldwork carried out since the early 1970s on the Triassic and Jurassic amphibian sites of Queensland by Dr Anne Warren, of La Trobe University, and her students has seen the discovery of many important new specimens, including an almost complete skeleton of one of the last-known labyrinthodonts, the Jurassic *Siderops*. Anne Warren has produced a number of notable students who have published on fossil

amphibians, such as Mark Hutchinson, Rob Jupp, Trevor Black, Ross Damiani, Adam Yates and Caroline Northwood. Triassic reptiles were also found in some of these sites, initiated by finds Bartholomai made in 1965–68. These were studied by him, Thulborn and Mr Tim Hamley. One of the bone scraps (from southern Queensland) was identified by Thulborn as a part of the skull of a mammal-like reptile, and recently several more bones have been found, representing the first fossil evidence of the therapsid group in Australia. Quite recently footprints attributed to this group have been described from New South Wales (Retallack 1996).

JOHN A. LONG

▲ Cape Paterson, Victoria, the site of one of Australia's earliest dinosaur finds

▼ The Dinosaur Cove site, Otway Ranges, Victoria
(Inset): Dr Tom Rich and Dr Pat Vickers-Rich, who have been working the Victorian dinosaur sites since 1979

MODERN DISCOVERIES IN VICTORIA

The rugged coastline of east Gippsland in Victoria has extensive outcrops of Lower Cretaceous rocks, which also occur to the west of Melbourne in the Otway Ranges. Although, as previously mentioned, a single dinosaur claw had been found near Cape Paterson (Woodward 1906), it was only in the late 1970s that a chance discovery of a dinosaur bone rekindled interest in the area. I still remember that day well, it was grey and blustery. Rob Glenie, a geologist, and my cousin, Tim Flannery, now at the Australian Museum, and I headed down to Cape Paterson to try and locate the site of the famous 'Cape Paterson claw'. Within minutes of reaching the rocky outcrops I picked up a small rock with the cross-section of black bone running through it! By the end of the day we had collected other fragments of bone and had demonstrated beyond doubt that the rocks there had the potential to yield bones.

Many other trips to the area by Flannery produced isolated dinosaur bones as well as other vertebrate remains. However, it wasn't until late

1979 to early 1980 that the Dinosaur Cove site, in the Otway Ranges, was discovered by Dr Tim Flannery and Dr Michael Archer. Here a rich lens of bone-bearing conglomerate was excavated over several years by Tom and Pat with the support of more than 500 volunteers. At times they faced great logistic problems and had to tunnel their way directly into the cliff face using rock drills, despite the violent waves crashing upon the cliff face. The sites on the east and west coast of Victoria have now produced over 5000 bones, with several new types of dinosaurs named (for example, *Leaelly-nasaura, Atlascopcosaurus, Timimus*); more material has also been uncovered of *Fulgurotherium*, a dinosaur described earlier from Lightning Ridge, New South Wales. In addition, scant remains of other dinosaur groups have been found, including unusual theropods and a possible neoceratopsian. The sites are still being worked and new surprises are turning up each field season. Recently an articulated leg of a small hypsilophodont dinosaur was recovered, as well as new material of a late-surviving labyrinthodont amphibian, including a skull.

Pat and Tom Rich have done much to bring dinosaurs into the public arena. In addition to their ongoing research work, they have produced several popular books on Australian prehistoric life, have been behind bringing several major exhibitions on dinosaurs and fossils to Australia, and were behind the idea of producing Australian dinosaurs on a series of postage stamps. These stamps were painted by artist Peter Trusler under the Richs' guidance, and were officially released in Australia on 1 October 1993. The sheet from which each stamp was cut comprises a scene showing not only well-known forms such as *Muttaburrasaurus* (75¢ stamp) and *Minmi* ($1.05 stamp), but new forms such as the ornithomimosaur *Timimus* (45¢ stamp), which at the time had not yet been published. The other animals on the stamp series include *Allosaurus, Leaellynasaura, Atlascopcosaurus* and the pterosaur *Ornithocheirus*. In September 1997 a further stamp series depicting Australian prehistoric animals was released. These reconstructions were done by Sydney artist Peter Schouten, and feature the dinosaur *Rhoetosaurus*, the plesiosaurian *Woolungasaurus* and the giant labyrinthodont *Paracyclotosaurus*.

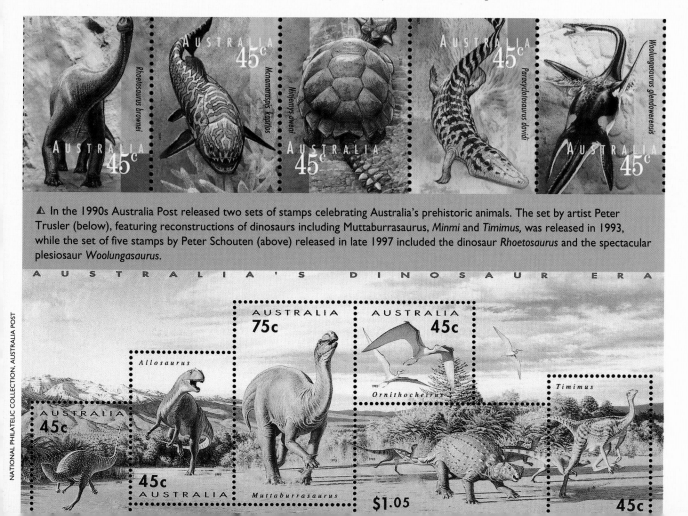

▲ In the 1990s Australia Post released two sets of stamps celebrating Australia's prehistoric animals. The set by artist Peter Trusler (below), featuring reconstructions of dinosaurs including Muttaburrasaurus, *Minmi* and *Timimus*, was released in 1993, while the set of five stamps by Peter Schouten (above) released in late 1997 included the dinosaur *Rhoetosaurus* and the spectacular plesiosaur *Woolungasaurus*.

MODERN DISCOVERIES IN SOUTH AUSTRALIA AND NEW SOUTH WALES

The extensive outcrops of Cretaceous rocks in central South Australia and northern New South Wales are well-known for their opal deposits. In these formations the bones dissolved away to be later replaced by opal (called 'pseudomorphs'). In 1932 the German scientist Friederich von Huene described opalised bones from Lightning Ridge, New South Wales, as belonging to a new hypsilophodont dinosaur, which he named *Fulgurotherium*, together with a tail vertebra he attributed to a new theropod, which he named *Walgettosuchus*. These bones, along with other more recent finds, were revised and described by Molnar (1980b), and by Molnar and Galton (1986).

The Coober Pedy opal field has produced several plesiosaur skeletons and isolated bones of dinosaurs, including the theropod dinosaur *Kakuru* described by Molnar and Pledge (1980). Unfortunately, because of their opal value and unique preservation, such specimens are often sold privately and do not find their way into museum collections. The *Kakuru* specimen suffered such a fate. It was first noticed by Neville Pledge, a palaeontologist from the South Australian Museum, who saw it on display in an Adelaide opal shop. The specimen was borrowed and a cast made, but shortly afterwards the bone was sold at auction and its whereabouts are not known. Other cases of plesiosaur skeletons being dug up and sold overseas have been discussed by Ritchie (1988).

Luckily, the famous opalised pliosaur skeleton nicknamed 'Eric' was obtained by the Australian Museum by tender when it came up for public auction. A public appeal, the brainchild of Dr Alex Ritchie, of the Australian Museum, raised sufficient funds to ensure that this beautiful specimen was secured for future generations to marvel at. The skeleton was meticulously prepared by hand over several hundred hours by a student, Paul Willis, when it was first brought in to the museum. The specimen is a nearly-complete skeleton with a well-preserved skull and also includes a last meal of gizzard stones and fish bones.

Opalised plesiosaur bones from White Cliffs in New South Wales were described as long ago as 1897 by Etheridge, but they disappeared to England shortly afterwards. In the mid-1980s a collection of fossil bones and plants from Lightning Ridge, New South Wales, was purchased by the Australian Museum on the strength of the presence of Australia's first Mesozoic mammal *(Steropodon)* being a part of the collection. Further searching in the area led Dr Alex Ritchie to the 'sheepyard site', which has so far been very productive at yielding many

▲ Dr Alex Ritchie, of the Australian Museum, searching for Triassic vertebrates in southern Queensland

▼ Dr John Cosgriff plaster jacketing a fossil amphibian bone in the Erskine Ranges of north Western Australia, about mid-1960s

small dinosaur bones (Ritchie 1988). Each year opal festivals are attended by palaeontologists in search of new fossil materials. This tactic paid off for Ritchie in 1994, when three additional Mesozoic mammal jaws were recognised. Despite only one having teeth, the finds have doubled the number of named Mesozoic mammals known from Australia, and quadrupled the actual number of specimens! As long as people continue to mine opal there is a good chance of other dinosaurs, marine reptiles and possibly mammals coming to light.

JOHN A. LONG

▲ Dr Tony Thulborn (*right*), Tim Hamley (*left*) and Duncan Friend (*behind*) at Gantheaume Point, Broome

▽ Greg Milner, John Long, Alex Ritchie, with a Western Australian Museum field party, excavating a pliosaurid skeleton near Kalbarri, Western Australia, in 1994

KRIS BRIMMELL

MODERN DISCOVERIES IN WESTERN AUSTRALIA

Fossil amphibian remains were first collected from the Erskine Range of the Kimberley district in 1953 by a field party from the Bureau of Mineral Resources, Canberra, and these finds led vertebrate palaeontologists back to the region in June 1960. That expedition comprised Dr David Ride, then Director of the Western Australian Museum; Dr Charles Camp and Dr John Cosgriff, both from the Department of Palaeontology, University of California, Berkeley; Dr Ken McKenzie from the University of Western Australia; and Mr Duncan Merrilees of the Western Australian Museum. During this trip the type skull of *Blinasaurus* was discovered by Dr Ride. Further material from the region was collected in July and August of 1963 by Dr Ride and Mr E. Carr, and by Dr E. Colbert of the American Museum of Natural History and Mr Merrilees in May 1964. In June and July of 1965 Prof. A. J. Marshall and Dr Jim Warren, from the Monash University Zoology Department, visited the area with Dr Cosgriff and collected a large amount of fossil bone material. Since then little further work has been done at these sites, but much has been published on the fossil material collected.

The extensive outcrops of Cretaceous rocks in the north of Western Australia have long been known for their dinosaur footprints, first recognised back in the 1940s and described by Western Australian Museum director Ludwig Glauert (1952). New finds by Mr Paul Foulkes, of Broome, and his friends, were initially studied by me and revealed at least six different types of dinosaur tracks. Ongoing research by Thulborn, Hamley and Foulkes is revealing even more footprint types, including some of the world's largest sauropod tracks.

The first pterosaur bone from Western Australia was recognised when I first started rummaging through the Western Australian Museum's fossil collections. It was found in 1960 near Exmouth but only recognised as a pterosaur in 1990 and described by Dr Chris Bennett, from the USA, and myself in 1991. The first dinosaur bones from the west were also described by me in 1992, with further scant remains of dinosaurs recognised in 1995–96. Field expeditions to the Kalbarri and Exmouth regions during 1991–94 uncovered the first articulated Mesozoic reptile skeletons in Western Australia (plesiosaurians and ichthyosaurs), prompted by initial discoveries made near Kalbarri by Glynn Ellis, Greg Molnar and Ian Copp, geology students at the University of Western Australia. In spite of its vast size, Western Australia has perhaps the poorest record of dinosaurs from anywhere in the country; the search continues, however, each year by Western Australian Museum field parties.

DISCOVERIES IN NEW ZEALAND

Marine reptile fossils were first announced from New Zealand in September 1861 by Sir Richard Owen at the meeting of the British Association for the Advancement of Science, at Manchester. The plesiosaur bones had been discovered from the Waipara River region in the South Island by a Scot, Thomas Hood Cockburn Hood in 1859. Hood took these bones to Sydney and showed them to William Clarke, a geologist, who

identified them as being from a *Plesiosaurus*; when he sent a cast of them to Frederick McCoy in Melbourne, however, they were identified as *Pliosaurus*. Hood later sent them to Owen and donated the specimens to the British Museum. Owen described them as a new species, *Plesiosaurus australis*. Many more discoveries were made in this area of the South Island by local collectors, who often donated them to the Canterbury Museum. Its director, Julius Haast, documented the discoveries and initiated a period of purposeful collecting to enrich his collections. One unfortunate incident at this time, though, was the loss of several fossil bones, including a partial mosasaur skull, when the ship, the *Matoaka*, set sail bound for England in May 1869, carrying fossil specimens for Sir Richard Owen, and was never seen again. Later that year Haast collected more fossil material from the Waipara River region.

In 1865 the Colonial Museum was established in Wellington, with James Hector as its director, and it also wanted collections of fossil reptile bones. A local collector, William Travers, had donated some fossil reptile bones to Hector in 1866, and in 1867 he and Travers visited the sites at Waipara. In 1872 Alexander McKay was employed by Haast to collect specimens for the Canterbury Museum. He soon sent in six large cases of fossil bones, including a partial skeleton of a plesiosaur, which was later described by Hutton in 1894 as a new species, *Cimoliasaurus caudalis*. McKay later collected fossil reptiles from Haumuri Bluff, from December 1872 through to February 1873, resulting in several tons of discoveries! He spent the winter of 1873 preparing the rock away from his finds. In his description of McKay's finds, Hector (1874) announced that he had found the remains of some 43 individual fossil reptiles belonging to at least 13 different species.

This was a great period in the history of vertebrate palaeontology in New Zealand, one that was largely the result of planned and deliberate collecting. Yet, one mystery remains unsolved. Collections made by McKay in 1874 and 1891 at Waipara, and in 1880–81 at Motunau, and Haumuri Bluff in 1876, include many tons of

PRINCIPAL MESOZOIC VERTEBRATE SITES IN NEW ZEALAND

Mangahouanga Stream (C)

NORTH ISLAND

C—Cretaceous
T—Triassic

Tinui (C)

Haumuri Bluff (C)

SOUTH ISLAND

Mt Potts (T)

Waipara River (C)

Shag Point (C)

Nugget Point (?T)

0   250
km

fine specimens: many of these cannot be traced today. According to the 13th annual report of the Colonial Museum, a collection of 'New Zealand fossil saurians' was sent to the famous American palaeontologist, Edward Drinker Cope, in December 1877, and another consignment of some 39 cases of bones was sent to him in 1889. Although Cope's receipt for the 39 cases was found in 1909, the fossils have never been located. This mystery was further investigated in 1970 with no outcome (Welles and Gregg 1971).

In November 1969 a further specimen of a partial skull and neck of a mosasaur was found by workers at the University of Canterbury Geology Department, and this was later described by Welles and Gregg as a new species, *Prognathodon waiparensis*.

The modern discoveries of Late Cretaceous fossil vertebrates in the North Island of New Zealand have been largely due to the hard work of a private fossil collector, Joan Wiffen, her colleagues W. Moisley and T. Crabtree, and her late husband, M. A. Wiffen. These amateur palaeontologists began combing the Mangahouanga

▲ Mr Thomas Hood Cockburn Hood, who collected the first Mesozoic reptile remains from New Zealand—at the mouth of the Waipara River in 1859

Stream beds for specimens in April 1973, and have laboriously prepared many of their finds out of the hard sandstone in home workshops. The first find of reptile bones from this site was made by geologist D. Haw in 1958 and this discovery led Wiffen and her friends to reinvestigate the site. In recent years Wiffen has taken on the task of describing the fossils, or contacting experts to work on them, and is thus primarily responsible for many of the finds outlined later (in chapter 8). The first record of a dinosaur bone from that country was published by Molnar as recently as 1981. The first New Zealand pterosaur was identified and published in 1988 by Wiffen and Molnar and they presented a major review of the known dinosaur and pterosaur finds from New Zealand in 1994 (Molnar & Wiffen 1994).

Dr Ewan Fordyce of the Geology Department, Otago University, has also been active in discovering new material. A superb plesiosaur skeleton was found at Shag Point, North Otago, and excavated by him with the help of volunteers in 1983. Since then he has made other finds of plesiosaurs and mosasaurs from that site.

◁ The famous Mangahouanga Stream site, North Island. Joan Wiffen (*right*) has found all of the dinosaur and pterosaur bones so far known from New Zealand

▷ Excavating fossil reptile bones from the Waipara Gorge, north of Canterbury. This is the site where the first Mesozoic reptile bones were uncovered in New Zealand. Dr Arthur Cruickshank is in the foreground

DR EWAN FORDYCE

DR EWAN FORDYCE

part two

AUST
and
NEW ZE
in
the TRIA

RALIA

ALAND

SSIC

The Triassic Period (245–208 million years ago) began after the retreat of seas from around the coasts of Australia, and well after the great ice-age from the Late Carboniferous through to the Early Permian. Most of the sediments laid down in the Triassic of Australia represent river and lake deposits, with only minor amounts of marine rocks. Australia's position at this time was in fairly low latitudes, probably with a humid, temperate climate, and marked seasonal rainfall variations. The land was covered with a great variety of plants—mostly ferns, horsetails, conifers, and seed ferns, especially the *Dicroidium* flora. Coal accumulated in swampy basins around the Sydney area, and much of the sandstone visible in the Sydney to Gosford region is of Triassic age. Many types of fishes and large labyrinthodont amphibians lived in the rivers and lakes. The fishes included many varieties of bony fish (osteichthyans) including the deep-bodied *Cleithrolepis,* as well as primitive sharks, some with large neck spines present (xenacanths). Fossil fish of Triassic age are well-known from the St Peters, Brookvale and Gosford areas in New South Wales, and from the Knocklofty Ranges near Hobart (Long 1991, 1995a).

The best-known amphibian and reptile sites of Triassic age in Australia are: in southern Queensland near Rolleston and Bluff, where the Arcadia and Rewan Formations crop out; in New South Wales the sandstones around Sydney (Narrabeen and Wianamatta Groups, and Hawkesbury Sandstone); in the north of Western Australia the Blina Shale outcrops of the Erskine Ranges, near Blina Station; in Tasmania the Knocklofty Formation sites around Hobart (Warren 1982). Only the Queensland sites and the Knocklofty Formation have produced some rare early reptiles; all others have just labyrinthodont amphibians and fishes. Fragments of a therapsid, or mammal-like reptile, have been recovered from Queensland, yet the rarity in Australia of these creatures is an anomaly, since elsewhere in Gondwana at this time they are very abundant. Australia has a diverse and interesting assemblage of Triassic amphibians, all being members of the temnospondyl group.

Although no dinosaurs have positively been recognised from the Triassic of Australia, some dinosaur bones from an

...eithrolepis, a Triassic freshwater fish from the Sydney district

...e Erskine Ranges, Kimberley district, Western Australia. Triassic
...ibian remains come from the Blina Shale, a poorly
...opping rock layer which sits below the prominent sandstone
...ng the flat-topped mesas

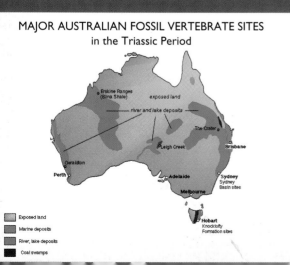

MAJOR AUSTRALIAN FOSSIL VERTEBRATE SITES
in the Triassic Period

Erskine Ranges
(Blina Shale)   exposed land
    river and lake deposits
        The Crater
    Leigh Creek   Brisbane
Geraldton
Perth   Adelaide   Sydney
                            Sydney
                            Basin sites
            Melbourne

                    Hobart
                    Knocklofty
                    Formation sites

Exposed land
Marine deposits
River, lake deposits
Coal swamps

unknown locality in north Queensland,
those of *Agrosaurus,* are thought to be of
Late Triassic or Early Jurassic age; so it is
also included in this chapter.

The Triassic of New Zealand has
yielded only a few fragmentary remains
of ichthyosaurs, and these have been
included at the end of this chapter.
During the Triassic the landmass that is
today New Zealand was still joined onto
western Antarctica, and would have had
similar climatic conditions to those of
southern Australia.

four

AMPHIBIANS AND REPTILES

AMPHIBIANS

CLASS AMPHIBIA
SUBCLASS LABYRINTHODONTIA
ORDER TEMNOSPONDYLI

SUPERFAMILY BRACHYOPOIDEA

The superfamily Brachyopoidea includes those temnospondyl amphibians with short, broad parabolic skulls, rounded orbits and nares, an absence of lacrimal bones, and a single anterior palatal vacuity. Occipital condyles extend well beyond skull table. Numerous other characters defining the group are listed by Warren and Damiani (1996).

FAMILY BRACHYOPIDAE

The family Brachyopidae contains most of Australia's fossil amphibians from the Triassic and Jurassic Periods. Brachyopids are characterised by having a short, broad skull with large orbits situated far forwards in the skull. The otic notch is either absent or very shallow; the tabular bones are short and broad and lacking tabular horns. Other anatomical details which characterise the family are listed and discussed by Chernin (1977), Warren and Hutchinson (1983), and by Warren and Black (1985), with the most recent diagnosis of the family being given by Warren and Damiani (1996).

GENUS BLINASAURUS

SPECIES *Blinasaurus henwoodi* Cosgriff 1968

AGE Lower Triassic

LOCALITY Erskine Ranges, about 90 km east of Derby, Western Australia (Blina Shale)

SPECIES *Blinasaurus townrowi* Cosgriff 1974

AGE Lower Triassic

LOCALITY Old Beach site, Derwent River, about 3 km north of Hobart, Tasmania (Knocklofty Formation)

SPECIES *Blinasaurus wilkinsoni* Stephens 1887

AGE Lower Triassic

LOCALITY Railway ballast quarry near Gosford, New South Wales (Narrabeen Group)

Blinasaurus henwoodi was described from an almost complete skull and lower jaw by Dr John Cosgriff, from Wayne State University, Michigan, USA in 1968. The name derives from the Blina Shale (named after Blina Station in the north of Western Australia), and from 'sauros' (Greek for 'lizard'), alluding to the lizard-like shape of such animals. The species name honours Mr John Henwood of Blina Station, who assisted the palaeontologists on their fieldwork in the Erskine Ranges. The skull of *Blinasaurus henwoodi* was about 12 cm long, giving an estimated overall size for the animal of about 60–70 cm. *Blinasaurus* probably fed chiefly on small fishes living in the ancient streams of the Kimberley.

▲ ▽ **S**kull of *Blinasaurus townrowi* from Hobart, viewed from (*above*), and showing the palate (*below*). (skull length 12 cm)

JOHN A. LONG

BLINASAURUS SPECIES FROM WESTERN AUSTRALIA, TASMANIA AND NEW SOUTH WALES Restorations of skulls

Blinasaurus townrowi from above (skull length 10 cm)

Blinasaurus townrowi the palate

Blinasaurus wilkinsoni
from above (skull length 3

TECHNICAL DATA Cosgriff (1967) used a differential diagnosis for the genus as he could not find any 'strictly definitive characters'. The head of *Blinasaurus* is rounded in top view, with large orbits and a flat, broad snout. Behind the eyes the skull deepens rapidly. Large fangs border the upper margin of the mouth. *Blinasaurus townrowi* is the best-preserved species, represented by a well-preserved skull and numerous isolated bones from the Old Beach site near Hobart. The skull is about the same size as for *B. henwoodi* but differs in several small details.

The very small skull of *B. wilkinsoni* (earlier named as *Platyceps wilkinsoni* by Stephens) probably represents a separate small species, rather than a juvenile: the orbits are proportionally smaller in

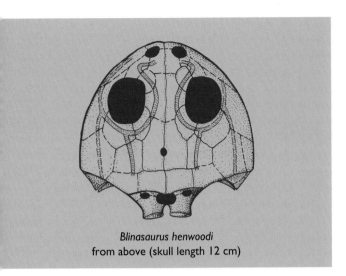

Blinasaurus henwoodi
from above (skull length 12 cm)

B. *wilkinsoni*, compared with *B. henwoodi*, and in juvenile labyrinthodonts the eye holes are usually larger than for adults. The total skull length of *B. wilkinsoni* is about 3 cm. Dr Anne Warren, of La Trobe University, has recently informed me that she believes that *Blinasaurus* may not be a valid genus, but could be synonymous with another well-known brachyopid, *Platycepsion*.

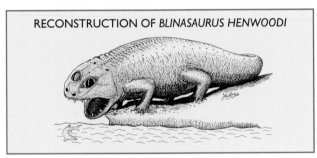

RECONSTRUCTION OF *BLINASAURUS HENWOODI*

⊿ Skull of *Xenobrachyops allos*, viewed from above (skull length 11 cm)

◁ Skull of *Blinasaurus henwoodi* from Western Australia (skull length 10 cm)

GENUS XENOBRACHYOPS

SPECIES *Xenobrachyops allos* Howie 1972a

AGE Lower Triassic

LOCALITY Near the headwaters of Duckworth Creek, southwest of the settlement of Bluff, south central Queensland

Brachyops was first described from remains found in Mangali, central India, by the famous anatomist and palaeontologist, Sir Richard Owen (1855). An Australian form similar to *Brachyops* was discovered from the erosion gullies near Duckworth Creek in southern Queensland, by Anne Warren and Alan Bartholomai during fieldwork in 1969 and 1970. The skull was found upside down with the front of the palate sticking out, suggesting that a whole skeleton might be present in the rocks. After digging a rather large hole around the skull, it was found that just the head was present. It was originally described as a new species of *Brachyops* by Howie (Anne Warren's maiden name) in 1972, but Warren and Hutchinson (1983) changed the name to *Xenobrachyops* after considering that the features seen on the Queensland species in fact distinguished it from the *Brachyops* known from India. The first part of the name *Xenobrachyops* comes from the Greek word for 'strange or foreign', alluding to the foreign nature of the '*Brachyops*' from Queensland. The skull is about 11 cm long and 14 cm across, and the mouth features several large fangs.

TECHNICAL DATA The head is broader than long, shallow anteriorly and deep posteriorly, as in other brachyopids; however, it differs from other brachyopids in the proportions of the skull and sizes of certain bones, and in the relative sizes of the interpterygoid vacuities of the palate. One of the primitive features of *Xenobrachyops* is the fact that the occipital condyles are almost level with the jaw articulation. In a recent review of brachyopid amphibians Warren and Damiani (1996) presented a new reconstruction of the skull roof and palate of *Xenobrachyops* and discussed its phylogenetic position, concluding that it and *Sinobrachyops*, from the Jurassic of China, are closely related and might even constitute their own separate family.

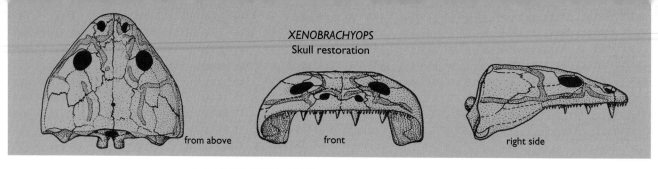

XENOBRACHYOPS
Skull restoration

from above   front   right side

GENUS NOTOBRACHYOPS

SPECIES *Notobrachyops picketti* Cosgriff 1973

AGE Upper Triassic

LOCALITY Hurstville Brick Company quarry at Mortdale, Sydney area (Ashfield Shale, Wianamatta Group)

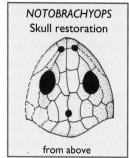

NOTOBRACHYOPS
Skull restoration

from above

Notobrachyops is known from a single impression of the skull roof and right lower jaw. The origin of the fossil is surrounded by mystery. The specimen was given to the New South Wales Geological Survey by Mr A. Garden. Shortly after Dr John Pickett took up the position of curator at the Survey Museum in 1965, he found the specimen in the collections and recognised it as an amphibian. It was sent to Dr John Cosgriff in Tasmania, who studied the specimen and named the species after Pickett (Cosgriff 1967, 1973). The genus name comes from 'notos' meaning 'southern' and the well-known genus *Brachyops*. It was a small animal, with a skull 35 mm in length, giving an estimated total length at about 30 cm.

TECHNICAL DATA *Notobrachyops* is a small brachyopid, characterised by having a narrow skull roof which is broadest across the quadratojugal bones, and the width is 0.9 times the midline length. Each skull roof bone is raised in its centre and depressed near its margins, giving an undulating appearance to the skull surface. Although Cosgriff (1973) placed *Notobrachyops* as being more closely related to other Australian brachyopids than to genera from elsewhere, Warren and Damiani (1996) excluded it from phylogenetic analysis because it is too incompletely known. Its Late Triassic age makes it one of the youngest of the known brachyopid labyrinthodonts.

FAMILY CAPITOSAURIDAE

Capitosaurs are characterised by having a long, broad snout with the orbits situated far back on the skull and the nostrils far forward; there is a well-developed otic notch situated near the rear corners of the skull, although this may close up in advanced forms. In overall appearance the head is somewhat crocodile-like.

GENUS PARACYCLOTOSAURUS

SPECIES *Paracyclotosaurus davidi* Watson 1958

AGE Middle Triassic

LOCALITY Brick pit, St. Peters, Sydney (Wianamatta Group)

Paracyclotosaurus is one of Australia's most spectacular fossil amphibians as it is known from a nearly complete skeleton 2.25 m in length and has a large skull armed with many small but sharp teeth. The skeleton was found by Mr B. Dunstan during the late 1800s inside a large ironstone nodule recovered from a brick pit. The name *Paracyclotosaurus* comes from its being very similar to a closely related beast called *Cyclotosaurus*. The

▼ Reconstruction of *Paracylotosaurus*, a large amphibian from the Sydney region by Peter Schouten

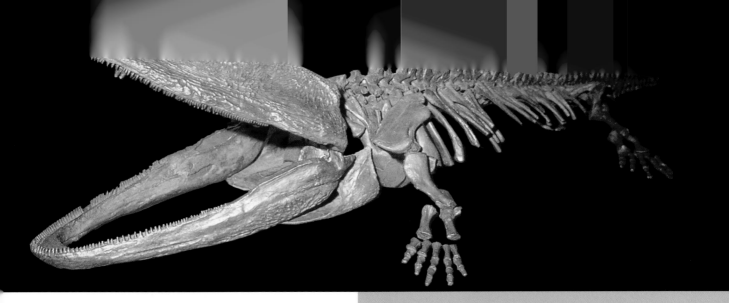

species name honours one of Australia's most eminent early geologists and Antarctic adventurers, Sir Edgeworth David. Fully reconstructed casts of the skeleton of this beast are on display in the Australian Museum, Sydney, and the Natural History Museum, London.

Impressions of the skin found with the skeleton indicate that the skin was dry, unlike that of modern amphibians, which use the skin as a supplementary breathing organ. However, the rib cage of *Paracyclotosaurus* suggests that the animal was better suited to breathing like a reptile. Much of the skeleton of *Paracyclotosaurus* was supported by cartilage, and the arms and legs were very small. *Paracyclotosaurus* probably spent most of its life in the water with only short trips onto land. Waiting in the freshwater lake for unsuspecting fishes to come near, it could then suck a fish into its large mouth simply by opening its jaws underwater. TECHNICAL DATA *Paracyclotosaurus* is a very large capitosaur, with a skull nearly 70 cm in length; the small orbits are situated in the posterior third of the skull, and the otic notch is closed. The tabulars are large, rounded bones, bigger than those in species of *Parotosuchus*, its close relative.

▲ Reconstructed skeleton of *Paracyclotosaurus davidi*, from a brick pit in Sydney

GENUS PAROTOSUCHUS

SPECIES *Parotosuchus brookvalensis* Watson 1956

AGE Middle Triassic

LOCALITY Brookvale, Sydney area (Hawkesbury Sandstone)

SPECIES *Parotosuchus wadei* Cosgriff 1972

AGE Middle Triassic

LOCALITY Sydney area (Narrabeen Group)

SPECIES *Parotosuchus gunganj* Warren 1980
Parotosuchus rewanensis Warren 1980

AGE Lower Triassic

LOCALITY The Crater, 72 km southwest of Rolleston, south central Queensland (Arcadia Formation, Rewan Group)

SPECIES *Parotosuchus aliciae* Warren and Hutchinson 1988

AGE Lower Triassic

LOCALITY Duckworth Creek, southwest of Rolleston, south central Queensland (Arcadia Formation, Rewan Group)

SPECIES *Parotosuchus* sp.

AGE Lower Triassic

LOCALITY Erskine Ranges, 90 km east of Derby, Western Australia (Blina Shale)

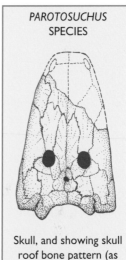

PAROTOSUCHUS SPECIES

Skull, and showing skull roof bone pattern (as preserved), from the Triassic Blina Shale, Western Australia:

(skull length 20 cm)

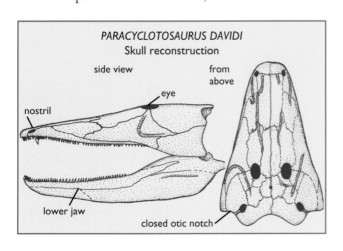

PARACYCLOTOSAURUS DAVIDI
Skull reconstruction

side view

from above

nostril

eye

lower jaw

closed otic notch

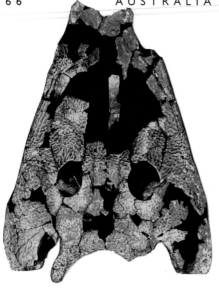

PAROTOSUCHUS SPECIES
Skulls from

DR ANNE WARREN

Parotosuchus is one of Australia's best-represented fossil amphibians, with species known throughout the country. The largest Australian species is *Parotosuchus gunganj*, which has a skull 24 cm in length, giving an overall estimated length for the animal of about 1.25 m. The species name comes from an Aboriginal word for water dweller (Warren 1980). The species *P. brookvalensis* is named after its locality, Brookvale, in Sydney. *P. wadei* is named after the Reverend T. Wade, who described many of the Triassic fish fossils of New South Wales.

Of particular interest are the small skulls of *P. aliciae*, named after Alice Crosland Hammersly, who found the juvenile skulls, since they show how *Parotosuchus* skulls change their shape and proportions with increased growth (Warren and Hutchinson 1988, 1990a; Warren and Schroeder 1995). Dr Anne Warren believes that the two Queensland species, *P. gunganj* and *P. rewanensis*, are more closely related to the South African and European species from the *Lystrosaurus* zone than to other Australian species, indicating the rapid dispersal of the species group throughout Gondwana.

TECHNICAL DATA The head of *Parotosuchus* is broad and flat with a blunt snout and small eyes situated far back on the top of the head. The genus is defined by its open otic notches, a single

◁ *Parotosuchus* sp. from the Triassic Blina Shale, Western Australia: skull, and showing skull roof bone pattern, as preserved

▽ Skull of *Parotosuchus aliciae*

DR ANNE WARREN

anterior palatal vacuity, and by both the frontal and jugal bones entering the orbits. The various species of *Parotosuchus* are distinguished by their relative proportions of the snout, dermal ornamentation and the relative degree of closure of the otic notch. *P. brookvalensis* has a semi-closed otic notch, which makes it the most advanced of the Australian species. The specimen from the Erskine Ranges, Western Australia, is not determinable as to species but is nonetheless known from a relatively complete skull 20 cm in length (Warren 1980). As it is in a hard ironstone rock and difficult to prepare without damaging the bone, the features of the palate have not been seen.

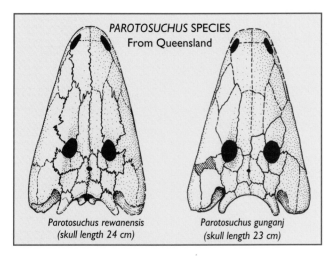

PAROTOSUCHUS SPECIES
From Queensland

Parotosuchus rewanensis
(skull length 24 cm)

Parotosuchus gunganj
(skull length 23 cm)

FAMILY CHUGITOSAURIDAE

The chugitosaurs were recognised as a distinct family in 1978, and are closely allied to the Brachyopidae by their overall shape and the features of the skull. Only a few genera are known: two from India *(Compsocerops, Kuttycephalus)*; the South American *Pelorocephalus*; the Australian *Keratobrachyops, Siderops,* and an as yet undescribed Victorian Cretaceous amphibian. Warren (1981b) characterises the family by a collection of proportional ranges and anatomical features, although none of these individually appear unique to the group. Chugitosaurids are similar to brachyopids in skull form but have well-developed tabular horns, smaller eye holes, many small teeth (rather than few larger teeth), and differ in a number of other anatomical features in the palate.

GENUS KERATOBRACHYOPS

SPECIES *Keratobrachyops australis* Warren 1981b

AGE Lower Triassic

LOCALITY Near the headwaters of Duckworth Creek, near the settlement of Bluff, south central Queensland (Arcadia Formation, Rewan Group)

Keratobrachyops was found in a similar fashion to other amphibian skulls from the Arcadia Formation—weathering out of the red siltstone. All the known specimens were found within about

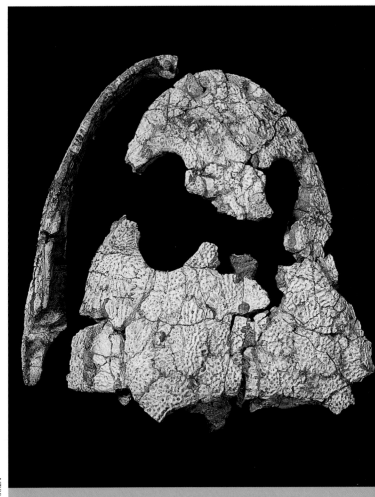

DR ANNE WARREN

▲ Skull of *Keratobrachyops*: viewed from above and showing the lower jaw (skull length 13 cm)

3 m of each other and only 1 m away from the type skull of *Brachyops allos*. The name comes from 'keras' the Greek word for 'horn', and alludes to the fact that it is the first 'horned' member (that is, having tabular horns) from the superfamily Brachyopoidea in Australia (hence the species name 'australis'). *Keratobrachyops* was recently redefined as a genus by Warren and Damiani (1996). It was a moderate-sized amphibian about 1 m long.

TECHNICAL DATA It has a narrow skull for a brachyopoid, slightly broader than long (breadth of the largest skull is 13.5 cm), and deeper than for

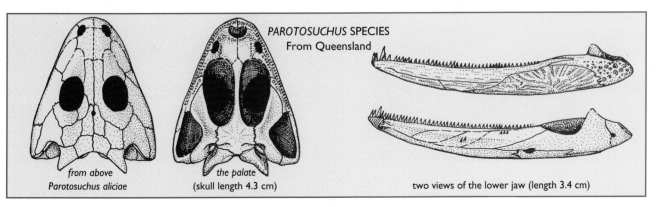

PAROTOSUCHUS SPECIES
From Queensland

from above
Parotosuchus aliciae

the palate
(skull length 4.3 cm)

two views of the lower jaw (length 3.4 cm)

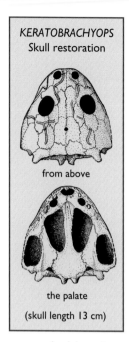

KERATOBRACHYOPS
Skull restoration

from above

the palate

(skull length 13 cm)

other chugitosaurids. The skull has the horns projecting from the tabular bones and the surface of the bone is covered with fine dermal ridges. The palate shows the presence of teeth and the absence of a palatoquadrate fissure, no contact between the exoccipital and the pterygoid, and the toothed maxilla bone enters the margin of the internal nostril (choana). It has been phylogenetically placed in two ways: either as a sister taxon to the Indian form *Kuttycephalus* (Sengupta 1995) or as a primitive sister taxon to the clade containing the South American form *Pelorocephalus*, plus the

two Indian genera and *Siderops* from the Jurassic of Queensland (Warren and Damiani 1996).

FAMILY LYDEKKERINIDAE

The Lydekkerinids are known from South Africa, Antarctica, and the one Australian genus. They are characterised by the slightly elongate shape of the skull, the centrally situated orbits, and by weakly developed sensory-grooves on the skull.

GENUS CHOMATOBATRACHUS

SPECIES *Chomatobatrachus halei* Cosgriff 1974

AGE Lower Triassic

LOCALITY Meadow Bank Dam site and Old Beach site, near Hobart (Knocklofty Formation)

Chomatobatrachus is represented by a well-preserved complete skull as the holotype specimen (from the Meadow Bank Dam site), as well as by several partial skulls, jaws and shoulder girdle

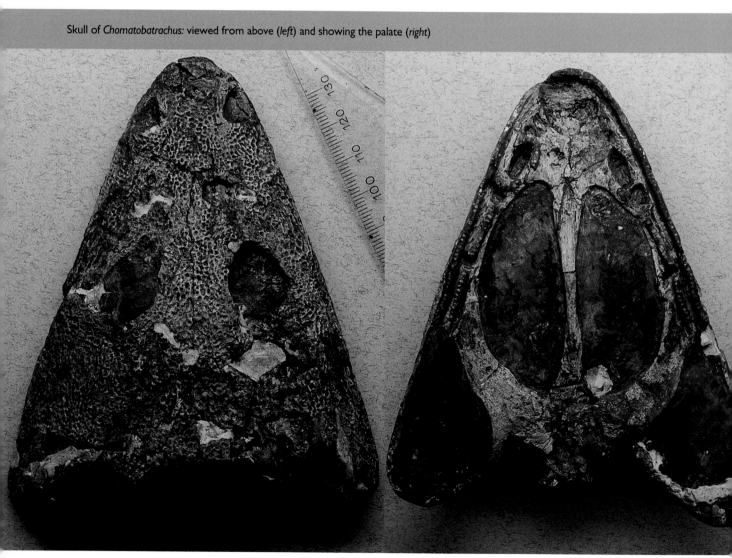

Skull of *Chomatobatrachus*: viewed from above (*left*) and showing the palate (*right*)

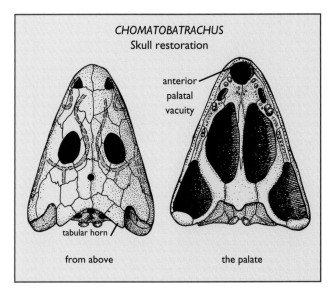

CHOMATOBATRACHUS
Skull restoration

anterior palatal vacuity

tabular horn

from above

the palate

bones from other sites around Hobart. The generic name comes from the Greek word 'chomatos', meaning 'mound' or 'rubbish heap', as the holotype skull came from rubble excavated during the construction of the Meadow Bank Dam. It was discovered by Mr Gordon Hale, then chief geologist for the Hydro-Electric Commission of Tasmania, and his name is honoured by the species. It was about 1 m in length.

TECHNICAL DATA The skull is about 11 cm in length, has an elongate shape with a slightly rounded snout, the sides of the skull are straight, and the palate has a circular anterior palatal vacuity. *Chomatobatrachus* is thought to be most closely related to *Lydekkerina* from South Africa (Cosgriff 1974).

FAMILY RHYTIDOSTEIDAE

Rhytidosteids are labyrinthodonts which are difficult to characterise in general terms. The skull is broad with a narrow, pointed front, and the eyes are situated in the front half of the skull. They are characterised by a number of anatomical details that readily separate them from other temnospondyl groups. Such details

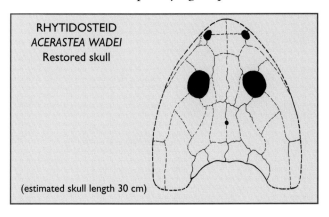

RHYTIDOSTEID
ACERASTEA WADEI
Restored skull

(estimated skull length 30 cm)

include having a short pterygoid-parasphenoid suture; exoccipitals not underplated by parasphenoid; exoccipital condyles with flattened elliptical articulating surfaces facing the midline; a narrow quadrate ramus of the pterygoid, a raised denticle field on the palatal series and a broad, flat cultriform process on the parasphenoid (Warren and Black 1985, Warren and Hutchinson 1987).

GENUS ACERASTEA

SPECIES *Acerastea wadeae* Warren & Hutchinson 1987

AGE Lower Triassic

LOCALITY The Crater, about 72 km southwest of Rolleston, south central Queensland. (Arcadia Formation, Rewan Group)

The type specimen is a poorly preserved partial skull with an estimated length of about 30 cm, giving it a size range of 1.5–2 m. The genus name means 'without horns', and the species name honours Dr Mary Wade of the Queensland Museum, who found the specimen only metres away from the sites where *Rewana*, *Parotosuchus rewanensis* and *P. gunganj* were all found. This specimen has been reconstructed from several large fragments, but nonetheless shows the general outline of the head and relative sizes of the eyes, and some associated body remains. Its importance lies in the presence of the associated body skeleton, a rare occurrence for the family. There were stones found with the skeleton that could be gastroliths (stomach grinding stones to aid digestion); alternatively, they may have been used for ballast to assist the animal when diving underwater.

TECHNICAL DATA The characteristic features of the genus are as follows: the deeply concave rear margin of the skull, the lack of tabular horns, the reduced post-temporal fenestra, and the presence of a flange transversing the posterolateral parts of the tabular and squamosal bones. The vertebrae have well-developed pleurocentra and elongated neural arches, and the scapulocoracoid is also well ossified. Ventral ribs covered the belly.

GENUS ARCADIA

SPECIES *Arcadia myriadens* Warren and Black 1985

AGE Lower Triassic

LOCALITY Near the headwaters of Duckworth Creek, southwest of the Bluff, south central Queensland. (Arcadia Formation, Rewan Group)

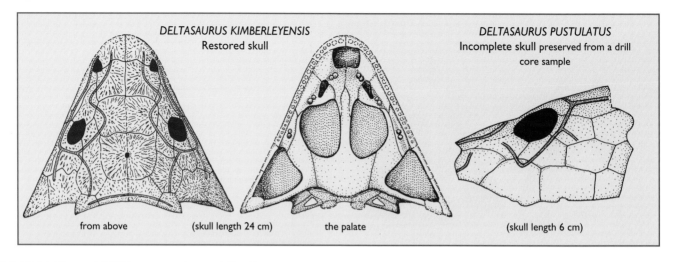

DELTASAURUS KIMBERLEYENSIS
Restored skull

DELTASAURUS PUSTULATUS
Incomplete skull preserved from a drill core sample

from above   (skull length 24 cm)   the palate   (skull length 6 cm)

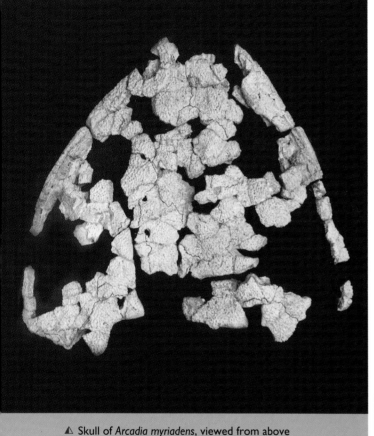

△ Skull of *Arcadia myriadens*, viewed from above

DR ANNE WARREN

The type specimen of *Arcadia* was discovered by Dr Gordon Sanson of the Monash University Zoology Department on a field trip with Anne Warren. *Arcadia* takes its name from the Arcadia Formation, and the species name alludes to the numerous small teeth it possesses. The skull, which measures 19 cm long and 17 cm wide, was turned over to Trevor Black, then a student at La Trobe University, who prepared and described it for his honours year thesis. This work led Warren and Black to regroup the temnospondyl amphibians and propose a new hypothesis on the relationships of Triassic temnospondyls.

TECHNICAL DATA The diagnostic features which separate it from other rhytidosteids are: the skull proportions (slightly longer than wide), together with the anterior position of the orbits; the unusually complex structure of the articular end of the lower jaw; the anterior palatal vacuity, with its little processes pointing towards the midline. Warren and Black (1985) suggested that *Arcadia* is closely related to the other Australian rhytidosteid *Rewana*, because of unique features seen in the lower jaw.

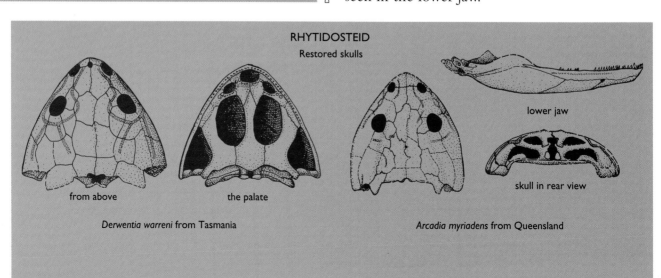

RHYTIDOSTEID
Restored skulls

lower jaw

skull in rear view

from above   the palate

Derwentia warreni from Tasmania

Arcadia myriadens from Queensland

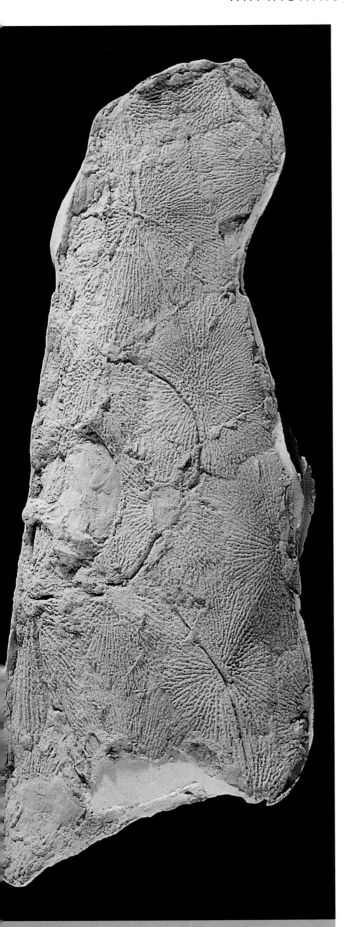

▲ Skull of *Deltasaurus kimberleyensis*, based on a latex peel of the impression

GENUS DELTASAURUS

SPECIES *Deltasaurus kimberleyensis* Cosgriff 1965

AGE Lower Triassic

LOCALITIES Erskine Ranges, about
90 km east of Derby, Western Australia (Blina Shale).
Also the following sites in Tasmania: Poatina Road,
near Launceston (Cluan Formation); Midway Point
and Conningham localities near Hobart (Knocklofty
Formation; Cosgriff 1974)

SPECIES *Deltasaurus pustulatus* Cosgriff 1965

AGE Lower Triassic

LOCALITY Bore hole near Geraldton, Western Australia
(Kockatea Shale)

Deltasaurus kimberleyensis was described from a complete left side of the skull, but numerous bones and fragments of the species have also been found, including well-preserved lower jaws. The name derives from the shape of the skull resembling the Greek letter 'delta' (Δ), the Greek for lizard ('sauros'), and the Kimberley district, where the fossils were found. Isolated bones and a partial skull of *D. kimberleyensis* were also described from the Tasmanian Triassic by Cosgriff (1974), giving this species a large geographical range, as also occurs with *Parotosuchus*. The skull of *Deltasaurus kimberleyensis* was about 24 cm long, giving an estimated size for the animal of close to 1.25 m, if the tail was relatively long.

A second, smaller species, *D. pustulatus,* was described from an imperfect skull recovered in a drill-core near Geraldton, Western Australia, at a depth of 806 m below the surface. The drill managed to core a nice section of the head, which included most of the skull roof and the right side of the cheek, thus providing enough of the head for it to be recognised as belonging to a new species. The chances of finding another skull in this manner, however, must be at least one in a million!

TECHNICAL DATA The head of *Deltasaurus kimberleyensis* is strongly triangular, moderately deep, and has small orbits; the pattern of bones on the skull roof is typical for labyrinthodonts. *D. pustulatus* differs from the Blina Shale species in being smaller in size, in having its bone surface ornamented with rows of wart-like pustules, and in the slightly narrower shape of its skull.

GENUS DERWENTIA

SPECIES *Derwentia warreni* Cosgriff 1974

AGE Lower Triassic

LOCALITY Old Beach site, on the Derwent River, and Midway Point site, near Hobart, Tasmania (Knocklofty Formation)

Derwentia takes its name from the Derwent River, which runs near the Old Beach amphibian fossil locality, and the species name honours the palaeontologist, Anne Warren. The holotype skull is about 9 cm long and is quite complete, apart from some abrasion to the edges suffered prior to burial. The teeth were small, indicating a creature of moderate predatory habits, probably a fish eater.

TECHNICAL DATA *Derwentia* is characterised by having large orbits situated in the front half of the skull; the skull roof is bumpy and irregular, with individual bones raised in their centres, and the exposed parts of the tabular bones are larger than exposed surfaces of the postparietal bones. *Derwentia* is an unusual rhytidosteid in having its eyes situated in the front half of the skull, and in the rounded sides of the skull, and the larger tabular horns. Cosgriff (1974) suggested it is closely related to *Deltasaurus* because of its similar skull shape and the development of the dermal ornament and lateral line canal grooves.

JOHN A. LONG

▲ Skull of *Derwentia warreni*, viewed from above (skull length 9 cm)

GENUS REWANA

SPECIES *Rewana quadricuneata* Howie 1972b

AGE Lower Triassic

LOCALITY The Crater, about 72 km southwest of Rolleston, south central Queensland. (Arcadia Formation, Rewan Group)

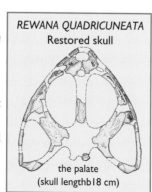

REWANA QUADRICUNEATA
Restored skull

the palate
(skull lengthb 18 cm)

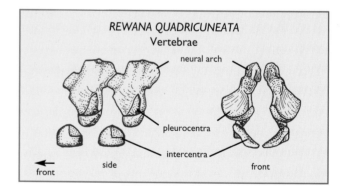

REWANA QUADRICUNEATA
Vertebrae

neural arch

pleurocentra

intercentra

front   side   front

Rewana was collected by a field party comprising Anne Warren, Alan Bartholomai, Alex Ritchie and Kingsley Gregg in 1969. The skull was found as weathered fragments of bones scattered over an area of 2 square metres; although some parts are missing, it is complete enough to show accurately the entire outline in palatal view. It measures about 18 cm estimated length. Some of the limb bones, vertebrae, ribs and fragments of the pelvic girdle were also found. The reconstructed animal would have been about 1 m long.

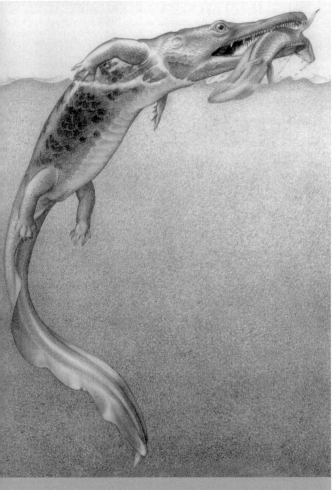

▲ Reconstruction of *Erythrobatrachus* by Peter Schouten

TECHNICAL DATA The skull is as broad as it is long, with small, rounded eye holes situated close to the skull's margins and far forward on the roof. The snout is slightly pointed, which distinguishes *Rewana* from other Australian broad-headed labyrinthodonts. The remarkable aspect of the vertebral column is that each backbone or vertebra has six parts: left and right sides of the neural arch, intercentra and pleurocentra. In most labyrinthodonts of the temnospondyl group the vertebrae have an intercentrum and two pleurocentra plus neural arches.

FAMILY TREMATOSAURIDAE

Trematosaurs are easily recognised by their long, narrow snout and narrow skull roof. The eyes are far back on the head and the nostrils open midway along the narrow snout.

GENUS ERYTHROBATRACHUS

SPECIES *Erythrobatrachus noonkanbahensis* Cosgriff and Garbutt 1972

AGE Lower Triassic

LOCALITY Noonkanbah Station, about 180 km east-southeast of Derby, Western Australia (Blina Shale)

Erythrobatrachus noonkanbahensis is the most easily recognisable of the Australian Triassic amphibians because of its long, pointed snout. Only parts of the skull are preserved, but enough to give an estimation of the shape of the head and pattern of bones between the eyes and nostrils and on the cheek, when compared with other trematosaurs. The genus name derives from two Greek words— 'erythros', meaning 'red', alluding to the colour of the fossil, and 'batrachos', meaning 'frog'. The species is named from Noonkanbah Station, in the Kimberley, where the holotype skull and another fragment were found within about 100 m of each other, in July 1960. Further specimens were found from the same locality in 1965. The estimated size of the complete skull is around 35 cm, making *Erythrobatrachus* one of the largest fossil amphibians from Western Australia, with an estimated maximum size close to 2 m. *Erythrobatrachus*, like all members of the trematosaur family, was a long-snouted fish-eater which lived in rivers and nearshore marine environments, probably occupying a similar niche to the saltwater crocodiles living today in northern Australia.

TECHNICAL DATA The genus is characterised by its small interpterygoid vacuities, and its short, broad skull roof proportions for the region bounded by the orbits, nares and lateral margins (as compared with the trematosaurids *Aphanerama* and *Wantzosaurus*). Another unique character of *Erythrobatrachus* is that the lateral margins of the skull bulge out around the large orbits.

GENUS INDETERMINATE

AGE Lower Triassic

LOCALITY The Crater, about 72 km southwest of Rolleston, south central Queensland (Arcadia Formation, Rewan Group); also, northeastern side of the Carnarvon Range near Moolayember Dip (Glenidal Formation, Clematis Group)

Warren (1985a) described some trematosaur snout fragments from two long-snouted amphibians, the first recorded from the Early Triassic of eastern Australia (Queensland), but their genus could not be determined. They represent an extension of the biogeographic range of the group within Australia.

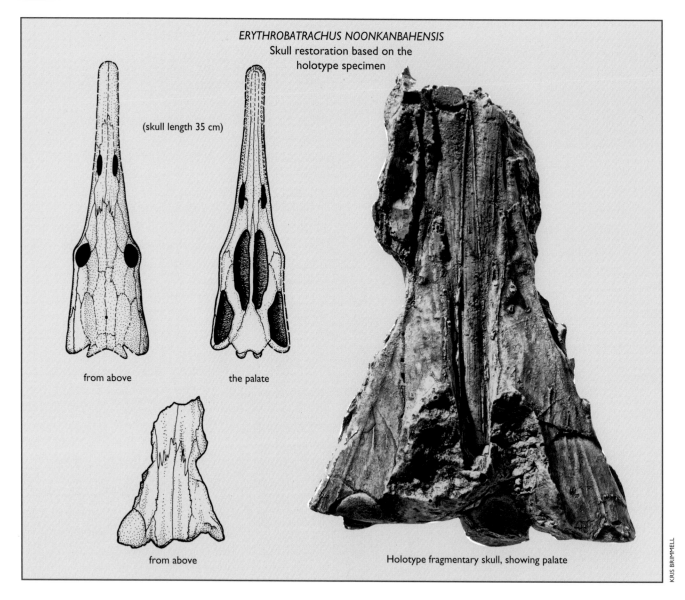

ERYTHROBATRACHUS NOONKANBAHENSIS
Skull restoration based on the
holotype specimen

(skull length 35 cm)

from above

the palate

from above

Holotype fragmentary skull, showing palate

KRIS BRIMMELL

FAMILY INDETERMINATE

GENUS LAPILLOPSIS

SPECIES *Lapillopsis nana* Warren and Hutchinson 1990a

AGE Lower Triassic

LOCALITY The Crater, about 72 km southwest of Rolleston, south central Queensland (Arcadia Formation, Rewan Group)

Lapillopsis is one of the most recently described Triassic amphibians from the famous Arcadia Formation of Queensland, a description based on very small skulls, less than 2 cm long. *Lapillopsis* does not easily fit into any of the main temnospondyl groups, but superficially resembles the micropholids. The name comes from a Latin word, 'lapillus', meaning 'pebble', and a Greek word, 'opsis', meaning 'appearance', as the tiny skulls were found in rocky nodules. The genus lacks sensory-line grooves and has a long-stemmed interclavicle bone underneath the shoulder region. These features suggest that the genus was more adapted for a terrestrial lifestyle than any of the other Triassic temnospondyls from the Arcadia Formation (Yates 1996).

TECHNICAL DATA Among the diagnostic features of the genus are the long jugal bone which extends in front of the orbits. As well, the skull has well-developed tabular horns and very large orbits, the latter probably being an immature feature. Sensory-line canals are not well-defined. *Lapillopsis* was thought to be closely related to *Micropholis*, from the Lower Triassic of Germany (Warren and Hutchinson 1990b) although Yates (1996) disagrees that the genus has any affinity to the dissorophoids but considers it to be more closely allied to stereospondyls.

FAMILY PLAGIOSAURIDAE

Plagiosaurs are characterised by their pustular ornamentation on bone surfaces, and the spool-shaped centra in the backbone. They are mainly known by well-preserved skeletons from Germany and Africa.

GENUS PLAGIOBATRACHUS

SPECIES *Plagiobatrachus australis* Warren 1985b

AGE Lower Triassic

LOCALITY The Crater, about 72 km southwest of Rolleston, south central Queensland (Arcadia Formation, Rewan Group)

The discovery of plagiosaurids in Australia was not made by a sudden find of a good skull, or any immediately recognisable bone, but rather by the slow process of becoming aware that some of the bone scraps collected from the Queensland Triassic vertebrate sites were unusual and could not be readily fitted into the existing faunal list. The name *Plagiobatrachus* comes from the Greek words for 'oblique' (referring to the family Plagiosauridae), and 'batrachos', meaning 'frog', with the species name alluding to Australia. Warren (1985b) believes that *Plagiobatrachus* is most closely related to *Plagiosuchus* from the Middle Triassic of Germany. It is difficult to estimate size from such fragments, but I estimate that the Australian plagiosaurid had a total body length of about 1–2 m.

TECHNICAL DATA The backbones are shaped like cotton-spools without a large central hole for the notochord. Fragments of lower jaw show that the surface ornament was very pustular, a distinct feature of the plagiosaurid group. These features, combined with the fact that the lower jaw piece was extremely flattened in life, enabled Warren (1985b) to diagnose this as a new genus of plagiosaurid.

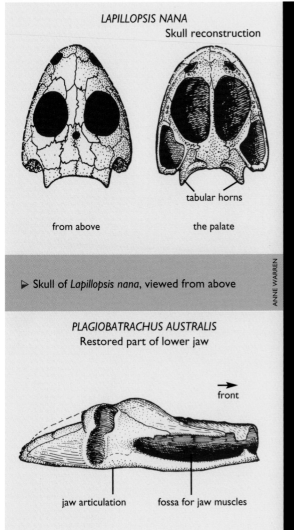

LAPILLOPSIS NANA
Skull reconstruction

tabular horns

from above

the palate

▷ Skull of *Lapillopsis nana*, viewed from above

ANNE WARREN

PLAGIOBATRACHUS AUSTRALIS
Restored part of lower jaw

front

jaw articulation

fossa for jaw muscles

REPTILES

Although Australia has no Triassic dinosaurs, the thecodonts, the group purported to have given rise to the dinosaurs, have been positively identified in Australia by *Kalisuchus* (from Queensland) and *Tasmaniosaurus* (from Tasmania). Thecodonts belong to the reptile group Archosauria, characterised by the specialised hand and ankle structure which preceded the advanced upright-walking type of ankle typifying all dinosaurs. The other Triassic reptiles known from Australia include *Kudnu*, a primitive lizard not very different from the surviving lizards, and *Kadimakara*, an eosuchian. Eosuchians have a diapsid type of skull with two large holes in each side, and include the earliest and most primitive members of the lizard-like reptiles (lepidosaurs). Characters used to diagnose the Lepidosauria and the Eosuchia have not been well-defined in the past and are currently under revision. The mammal-like reptiles which predominate other Gondwana faunas of this age are almost absent from Australia, apart from some fragments of bone and a set of footprints. Dinosaur footprints are known from the Late Triassic of Queensland, and one dinosaur, *Agrosaurus*, is of possible Triassic age and so is included here.

▲ Late Triassic ichthyosaur site in the Otamita Stream, New Zealand

▼ Triassic ichthyosaur vertebrae as weathering out, from the Southland district of New Zealand

Two reliable occurrences of ichthyosaur remains are known from the Late Triassic of New Zealand. The oldest-known Triassic ichthyosaur from New Zealand is based on some vertebrae found in the Torlesse Supergroup (Late Triassic, Oretian Stage) near Mt Potts, central Canterbury, South Island. These were described by Hector (1874) as belonging to 'Ichthyosaurus australis'. A later study of the specimens by Fleming et al. (1971) concluded that the remains were indeterminate as to genus and species. The second ichthyosaur comprises some broken jaws and teeth found at Otamita Stream, Hokonui Hills, Southland (Otamitian Stage, Late Triassic). These were described and figured by Campbell (1965), and comments by British palaeontologist Dr Alan Charig, published within that paper, suggest that the specimen has some characteristics of the family Shastasauridae.

Other Triassic vertebrates from New Zealand include mention by Fleming et al. (1971) of teeth with 'labyrinthodont characters' from unspecified formations near South Otago and Nelson. Fordyce (1991) has commented that these specimens were never properly described and their current whereabouts are unknown. They could have represented either ichthyosaur or labyrinthodont teeth. Fordyce (1991) also mentions a report of an indeterminate ichthyosaur from near Nelson, but no further details are provided.

CLASS REPTILIA
SUBCLASS LEPIDOSAURIA
ORDER SQUAMATA

FAMILY ?PALIGUANIDAE
GENUS KUDNU

SPECIES *Kudnu mackinlayi* Bartholomai 1979

AGE Lower Triassic

LOCALITY The Crater, 72 km southwest of Rolleston, south central Queensland (Arcadia Formation, Rewan Group)

Like *Kadimakara*, *Kudnu* is known only from a small partial skull preserved in a red mudstone nodule. The front of the snout and back of the skull are missing, but enough is preserved to allow comparisons with lizards. The overall size of *Kudnu* is estimated at about 20 cm, and in appearance it probably would not have been distinguishable from any typical living lizard. The name *Kudnu* comes from an Aboriginal mythical name for the 'lizard man', and the species name honours Mr R. McKinley, who assisted field parties working on the Rewan property.

TECHNICAL DATA *Kudnu* has a lacrymal bone situated between the nasal and maxilla, thus demonstrating that it is a true lizard. Originally it was thought to be closely related to *Paliguana* from South Africa (Bartholomai 1979), although Molnar (1991) regards it as too incomplete for confident assignment to any specific lizard family.

ORDER
EOSUCHIA

FAMILY PROLACERTIDAE
The prolacertids are not a well-defined group of reptiles. They are lizard-like in overall form, with hind limbs slightly longer than the arms for short bursts of bipedal running.

DR EWAN FORDYCE

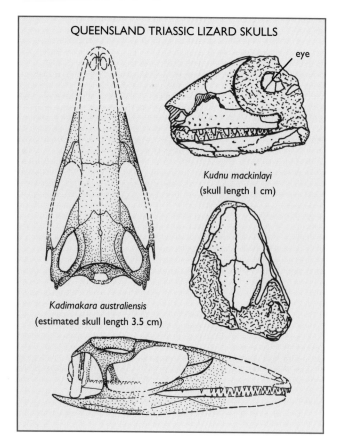

QUEENSLAND TRIASSIC LIZARD SKULLS

eye

Kudnu mackinlayi
(skull length 1 cm)

Kadimakara australiensis
(estimated skull length 3.5 cm)

GENUS KADIMAKARA

SPECIES *Kadimakara australiensis* Bartholomai 1979

AGE Lower Triassic

LOCALITY The Crater, 72 km southwest of Rolleston, south central Queensland (Arcadia Formation, Rewan Group)

The single incomplete skull of *Kadimakara* shows that the animal would have been about 35 cm long, by comparison with the African *Prolacerta*, which it closely resembles. The name *Kadimakara* comes from Aboriginal words for 'animals of the Dreamtime', with the species name indicating it is from Australia (Bartholomai 1979). Eosuchians like *Kadimakara* and *Prolacerta* had larger hind legs for bursts of upright running, a feature which became the hallmark of advanced archosaurians, and ultimately led to the rise of dinosaurs. Molnar (1991) suggests that *Kadimakara* was probably an insect-eater.

TECHNICAL DATA The diagnostic features of *Kadimakara* are that the skull shows the absence of a lower temporal bar and the way the lachrymal bone tapers out at the nasal bone, separating the prefrontal and maxilla bones. Molnar (1991) examined the specimen to see if it possessed features recently defined by Benton (1985) for the prolacertids. Molnar found *Kadimakara* possesses a tetraradiate squamosal and that there is a gap between the pterygoids to accommodate the parasphenoid. These characters indicate it is in accord with Benton's definition of the group.

SUBCLASS ARCHOSAURIA
ORDER THECODONTIA

FAMILY PROTEROSUCHIDAE

Proterosuchids are recognised by, amongst other features, the presence of triple-headed ribs, and teeth which are not fully anchored to the jaw bone (subthecodontian type). They were mostly aquatic, predaceous reptiles living near ponds, using swimming as their main form of locomotion (Benton 1979).

GENUS KALISUCHUS

SPECIES *Kalisuchus rewanensis* Thulborn 1979

AGE Lower Triassic

LOCALITY The Crater, 72 km southwest of Rolleston, south central Queensland; also some bones from near Duckworth Creek, 127 km north-northeast of the Crater (Arcadia Formation, Rewan Group)

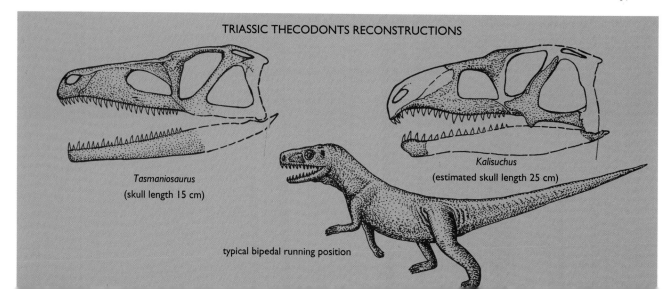

TRIASSIC THECODONTS RECONSTRUCTIONS

Tasmaniosaurus
(skull length 15 cm)

Kalisuchus
(estimated skull length 25 cm)

typical bipedal running position

Kalisuchus is represented by a variety of isolated and fragmentary skeletal remains including skull bones, vertebrae, limb and girdle bones. The name derives from Kali, the Hindu goddess associated with destruction, alluding to the fact that the skeleton was found in many small fragments, and from the Greek word 'suchus', meaning 'crocodile'. It is an important genus because it is the oldest archosaur known from Australia. Archosaurs are the group containing thecodonts, dinosaurs, crocodilians, and birds. The skull is represented by a jugal bone, and a partial skull roof as well as fragments of the upper and lower jaws. The limb bones show that one of the ankle bones (calcaneum) is strikingly like that of a crocodile. *Kalisuchus* was thought to be an amphibious, crocodile-like predator which is closely related to the well-known Chinese and African genus *Chasmatosaurus* (Thulborn 1979). It had an estimated maximum body length of 3 m.

TECHNICAL DATA *Kalisuchus* is characterised amongst thecodonts by its crocodiloid calc-aneum in the ankle, and the upper jaw shows the presence of a maxillary shelf, suggestive of a broad snout. Its limb bones are slender, and the neck is relatively long, for a proterosuchian.

GENUS TASMANIOSAURUS

SPECIES *Tasmaniosaurus triassicus* Camp and Banks 1978

AGE Early Triassic

LOCALITY Crisp and Gunn's quarry at the head of Arthur Street, West Hobart, Tasmania (Poets Road member of the Knocklofty Formation)

Tasmaniosaurus is Australia's most complete Triassic reptile—for that matter, it is the most complete fossil reptile known in Australia, as the whole skeleton is preserved as the scattered remains of a single individual on a slab of sandstone. The specimen was found in September 1960 by Dr J. Townrow and Dr Max Banks who saw the bones in a loose block of sandstone at Crisp and Gunn's quarry, Hobart. *Tasmaniosaurus* was first described by Camp and Banks (1978) and later redescribed by Thulborn (1986). The animal was lizard-like in appearance and about 1 m in total length. Some small bone fragments of a labyrinthodont amphibian are found associated with the ribs of *Tasmaniosaurus*, possibly indicating what may have been its last meal. Thulborn (1986) believes that

Tasmaniosaurus is similar to, and possibly a close relative of *Chasmatosaurus*, from the Triassic of Africa and China.

TECHNICAL DATA Diagnostic features are: the long, slightly curved premaxilla, the absence of parietal foramen, vertical quadrate, vacuity at posterior end of dentary, shallowly amphi-coelous vertebrae, long double-headed cervical ribs, long limbs and feet, and the absence of bony scutes (Camps and Banks 1978). Relative to all other proterosuchians, *Tasmaniosaurus* is primitive in being the only member of the group which retains an interclavicle. The skull features a narrow snout and the antorbital fenestra has a rounded anterior margin, unlike the square-shaped one in *Kalisuchus*, and the premaxilla has an unusually large number of teeth (about 16)

ORDER SAURISCHIA

FAMILY PROSAUROPODIDAE

Prosauropods were long-necked, long-tailed, mostly upright, walking dinosaurs. They are the ancestral group that gave rise to the great sauropods, including well-known forms like *Brachiosaurus* or *Diplodocus*. Prosauropods occupied the niche of large omnivorous dinosaurs of the Triassic, some being flesh-eaters but others developing plant-eating dentations.

GENUS AGROSAURUS

SPECIES *Agrosaurus macgillivrayi* Seely 1891

AGE Late Triassic–Early Jurassic

LOCALITY northern Queensland coast, possibly near the tip of Cape York (Molnar 1991)

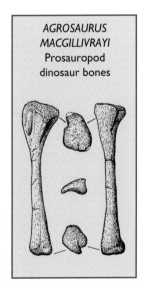

AGROSAURUS MACGILLIVRAYI
Prosauropod dinosaur bones

The mystery surrounding the discovery of these bones by crew members of HMS *Fly* has been discussed earlier (chapter 3); it need only be repeated here that it is uncertain what the exact location the bones were found in is, or what the age of the specimens is, or whether they even came from Australia (Rich *et al*, 1998). *Agrosaurus* is known from a

▲ Bones (close to actual size) of the prosauropod dinosaur, *Agrosaurus macgillvirayi*, from an unknown location, possibly in northeastern Queensland

JOHN A. LONG

ORDER THERAPSIDA
SUBORDER DICYNODONTIA

FAMILY ?LYSTROSAURIDAE

GENUS INDETERMINATE

SPECIES Indeterminate

AGE Early Triassic

LOCALITY The Crater, about 72 km southwest of Rolleston, south central Queensland (Arcadia Formation, Rewan Group)

The discovery of a single, small bone fragment from the Triassic rocks exposed in the Crater site, in the early 1980s, led Dr Tony Thulborn (1983) to recognise the first definite occurrence of the mammal-like reptiles, or therapsids, from Australia. This group is abundant in other Gondwana Triassic localities, and its absence from Australia had long been a mystery. This bone, which is part of the upper jaw joint (quadrate), is particularly characteristic in its shape for some of the mammal-like reptiles, enabling identification of the Australian form as a close relative of *Lystrosaurus* from the Triassic of South Africa. Other therapsid bones from the same locality include part of a large canine tusk, two small vertebrae, and another skull fragment. The latter is interesting but requires further preparation before it can be properly studied (Thulborn 1990b). The dicynodonts, to which the bone fragments belong, were plant-eating reptiles with enlarged canine tusks that were probably used for digging for roots, defending themselves, or possibly for mating rituals. They were stout, heavily-built animals, whose average size was comparable to that of a sheep.

TECHNICAL DATA In dicynodonts the quadrate bone is developed as a double knuckle so that the lower jaw articulated to it almost in pulley-like fashion.

well-preserved shin bone (tibia), a claw, and some other fragments collected by the ship's crew in 1844 and taken back to England. When further preparation of the original block of matrix containing the bones was carried out by staff at the Natural History Museum in London in the late 1980s, some further smaller bones, including a small vertebra, were uncovered. In their review of the anchisaurid prosauropods Galton and Cluver (1976) pointed out that *Agrosaurus* is a prosauropod dinosaur similar to *Anchisaurus*, and Molnar (1991) adds that it is probably very close to *Massospondylus* from Southern Africa.

TECHNICAL DATA *Agrosaurus* is characterised amongst prosauropods by the features of its tibia. The dorsal surface is much longer antero-posteriorly than across its anterior breadth, the distal end has a well-defined lateral notch for the astragalus.

PRIMITIVE PROSAUROPOD DINOSAUR
RECONSTRUCTION
Indicating the possible appearance of *Agrosaurus*

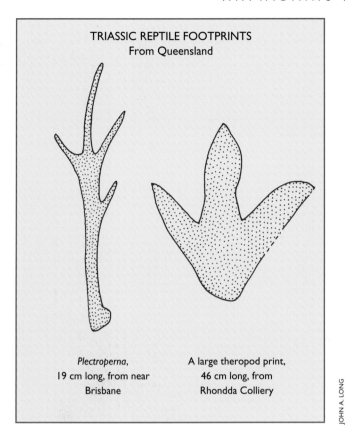

TRIASSIC REPTILE FOOTPRINTS
From Queensland

Plectroperna,
19 cm long, from near
Brisbane

A large theropod print,
46 cm long, from
Rhondda Colliery

JOHN A. LONG

▲ Skull of *Lystrosaurus* (from South Africa), a very similar animal to the Queensland therapsid

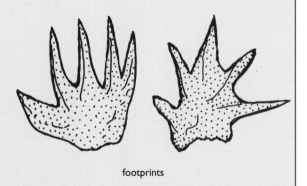

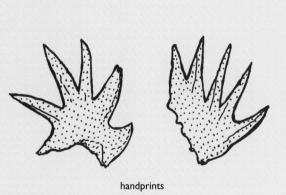

THERAPSID REPTILE
from Bellambi Colliery, near Wollongong in
New South Wales, named as
Dicynodontipus bellambiensis

footprints

handprints

(after Retallack 1996)

REPTILE FOOTPRINTS

Dr Alan Bartholomai (1966a) has reported large dinosaur footprints from the Rhondda Colliery near Dinmore, southeastern Queensland, of Late Triassic age (Blackstone Formation). These are large three-toed tracks up to 46 cm in length, with a stride of 2 m, and belong to a carnivorous theropod dinosaur whose estimated length is about 6 m. Bartholomai has compared these to tracks described as *Eubrontes* from the Connecticut Valley of North America.

A second type of reptile track has been found at Bergin Hill Quarry, near Goodna, close to Brisbane. Molnar (1982b) identified these as *Plectroperna* sp., a conclusion based on comparisons with two species of *Plectroperna* from the Portland Arkose of Connecticut and Massachusetts, North America. The longest of the Queensland footprints is 19 cm long, and they differ from the ones in America in being larger, more slender, and with the first toe print (hallux) pointed to the front. They probably belonged to either a thecodont or lizard-like beast. Unstudied reptile footprints are also known from the Triassic rocks of the Sydney Basin, at Berowra Creek (Fletcher 1948).

DR ANNE WARREN

◄ Reconstruction by Peter Schouten of a mammal-like reptile, similar to the one that inhabited Australia in the Triassic Period, currently known by only a few scant remains.

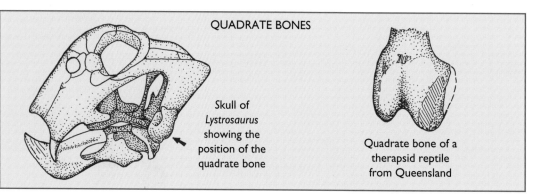

QUADRATE BONES

Skull of *Lystrosaurus* showing the position of the quadrate bone

Quadrate bone of a therapsid reptile from Queensland

▲ The Crater site, Queensland, showing red mudstone layers that contain the fossilised remains of many species of Early Triassic amphibians and reptiles

Most recently Retallack (1996) described well-preserved fossil trackways belonging to a therapsid, possibly *Lystrosaurus*-like animals. They include associated pentadactyl handprints and footprints, and are of Early Triassic age from the Bellambi Colliery in the southern Sydney basin. The tracks were described as *Dicynodontipus bellambiensis* (Retallack 1996) and indicate that the animal making the tracks was between 84 cm and 1 m long, intermediate in size for *Lystrosaurus*, one of the commonest Early Triassic therapsids in Gondwana countries. The animals walked with a gait that was more reptilian than mammalian, and were believed to be slowly approaching and returning from water across an open floodway between the receding river and a forested

The Jurassic Period (208–144 million years ago) saw freshwater sediments with shallow marine incursions accumulate along the western margin of Australia, near Geraldton, and in the Kimberley districts, with a large, inland river system developing in central Queensland and northern New South Wales. Freshwater river and lake sediments deposited in this large drainage basin have produced fish fossils at several localities (Long 1991, 1995a), but the bones of reptiles and amphibians are very rare. The climate in Western Australia was thought to be warm, based on oxygen isotope studies of marine sediments; but eastern Australia appears to have been cooler with abundant rain, and coal swamps forming in southern Queensland. The land was covered with many varieties of plants, but dominated by cold-weather forms, including forests of conifers, with abundant ferns, cycads and seed ferns forming the undergrowth. Fossil insects such as beetles occur in the Middle Jurassic of Western Australia, near Geraldton.

During the Jurassic Period Gondwana underwent radical changes as the major continents began to break away. Africa and South America drifted away from Antarctica, and India began its long journey northwards towards Asia. By the close of the Jurassic Period Australia was left joined to Antarctica. These tectonic movements had profound influences on changing global climates as sea currents altered their courses. The nature of Jurassic vegetation in Australia has remained almost unchanged in remote locations in Queensland, where isolated forests of conifers like *Araucaria* still exist.

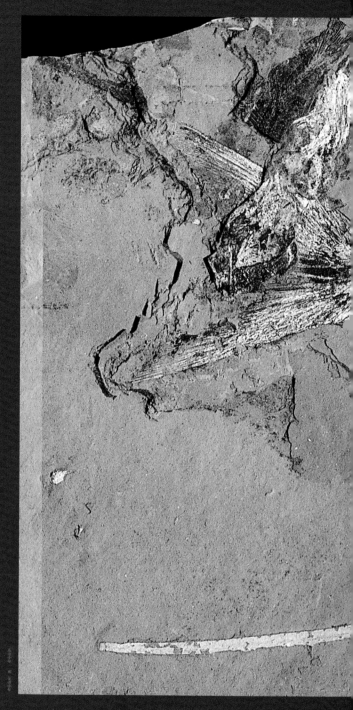

Although Australia has abundant Jurassic invertebrates, land animals of this age are very rare. Well-preserved fish fossils from Talbragar River in New South Wales are typical of Jurassic faunas from throughout the world, and include early representatives of the teleostean bony fishes, such as *Leptolepis*, as well as relict primitive forms like the palaeoniscoids (Long 1995a). The few land animals known from this age mostly

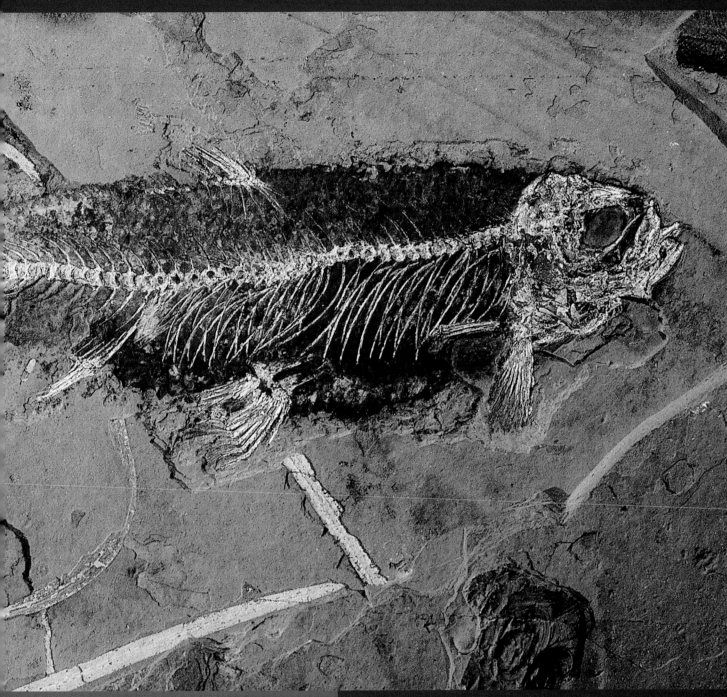

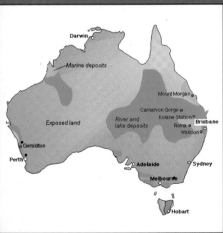

come from Queensland, with a few fragments from Western Australia. The faunas include the large sauropod dinosaur *Rhoetosaurus*, a fragment of theropod dinosaur, some aquatic reptiles such as plesiosaurs, and one of the last-known labyrinthodont amphibians, *Siderops*. Footprints from Queensland indicate that several other dinosaur types were living in Australia at the time, including very large, flesh-eating forms.

five

AMPHIBIANS, MARINE REPTILES AND DINOSAURS

AMPHIBIANS

CLASS AMPHIBIA
SUBCLASS LABYRINTHODONTIA
ORDER TEMNOSPONDYLI

SUPERFAMILY BRACHYOPOIDEA

FAMILY CHUGITOSAURIDAE

GENUS SIDEROPS

SPECIES *Siderops kehli* Warren and Hutchinson 1983

AGE Uppermost Early Jurassic

LOCALITY Kolane Station, south central Queensland
(Westgrove Ironstone member of the
Evergreen Formation)

The discovery of *Siderops*, known from an almost-complete skeleton, caused a palaeontological sensation. Not only was it one of Australia's largest (2.5 m long) and most complete fossil amphibians, but it came from rocks of dated upper Early Jurassic age—a time when labyrinthodonts were long thought to be extinct. Thus, *Siderops* is truly a living fossil for its age, as most other brachyopoids had died out by the Late Triassic, some 30 million years earlier! Since the discovery of *Siderops*, there has also been a Jurassic labyrinthodont discovered from China (*Sinobrachyops*), and just recently a labyrinthodont has been found from the Early Cretaceous of Victoria, indicating the group survived longer on this continent than anywhere else. The name *Siderops* comes from Greek words meaning 'iron' and 'face', alluding to the hard ironstone rock the bones were found in. Preparation of the find took several years of hard work to remove the rock from the skeleton before it could be reconstructed. The species name honours the Kehl family, who found the skeleton.

Siderops has a large skull, which is broader relative to its body than other known labyrinthodonts. The head is massive and the arms and legs are quite small, despite which it was capable of occasionally moving out of the water onto land. It was armed with many small, sharp teeth, which suggests it fed on fishes and other small animals, and could have caught them simply by lying in wait for its prey to come near and then opening its massive mouth and sucking the prey inside. The large mouth would also have been capable of catching and killing small land animals that unsuspectingly came too close to *Siderops*. An

interesting find was made in the rock inside the mouth of *Siderops*—a small fossil millipede, *Decorotergum*, the only known Mesozoic record of these arthropods (Jell 1983). TECHNICAL DATA *Siderops* is characterised by its extremely broad head, which has extended bony horns on the tabular bone. Recent analyses have placed *Siderops* as one of the most advanced members of the family, closely allied to *Compsocerops* from India (Warren and Damiani 1996, Sengupta 1995).

Siderops, reconstruction by Brian Choo

FAMILY ?BRACHYOPIDAE
GENUS AUSTROPELOR

SPECIES *Austropelor wadleyi* Longman 1941

AGE Uppermost Early Jurassic

LOCALITY Bed of the Brisbane River near Lowood Station, south central Queensland (Marburg Sandstone)

When the fossil jaw of *Austropelor* was first studied, its age was uncertain. A suggestion that it was Jurassic age was thought unlikely as labyrinthodonts were then not known to occur in Jurassic rocks, so the jaw was deemed to be reworked from older Triassic rocks. Since the discovery of *Siderops*—of undisputed Jurassic age—and, furthermore, of Cretaceous amphibians in Victoria, it has now been recognised that *Austropelor* is also from the Jurassic.

TECHNICAL DATA *Austropelor* is known from one fragmentary lower jaw, and hence its assignment to a separate genus is not certain. Warren and Hutchinson (1983) conclude that it is similar to the jaw of *Siderops*, but nothing precludes it being assigned to either the brachyopids or chugitosaurids.

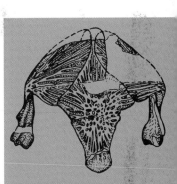

▲ Shoulder girdle bones of *Siderops kehli*, viewed from underneath

▼ Fragmentary lower jaw of *Austropelor wadleyi* from Queensland,: viewed from the side, and from above

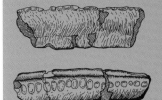

▲ Skull of *Siderops kehli* from the Early Jurassic of Queensland (approximate skull length 50 cm)

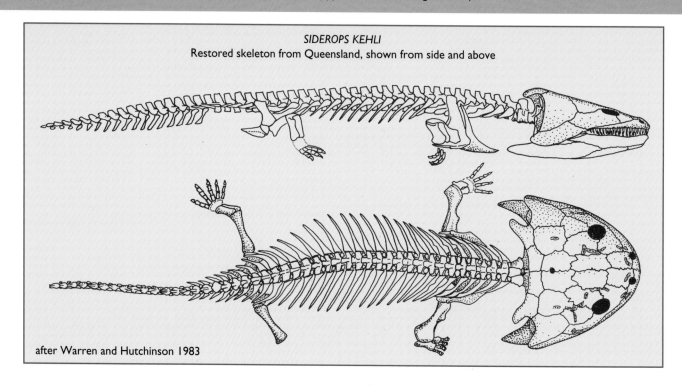

SIDEROPS KEHLI
Restored skeleton from Queensland, shown from side and above

after Warren and Hutchinson 1983

MARINE REPTILES

The only other Jurassic reptiles known from Australia are the plesiosaurians, aquatic reptiles which lived in rivers and shallow seas and mostly fed on fish. The few remains of Australian Jurassic plesiosaurians come from Queensland and Western Australia.

SUBCLASS SAUROPTERYGIA
ORDER PLESIOSAURIA

SUPERFAMILY PLIOSAUROIDEA

FAMILY ?PLIOSAURIDAE

Pliosaurids were relatively short-necked, long-snouted, robust plesiosaurians. I follow Brown (1981) here in lumping all of the pliosaurs in one family, rather than including *Leptocleidus* in the separate family Rhomaleosauridae (Molnar 1982b).

GENUS INDETERMINATE

SPECIES Indeterminate

AGE Early Jurassic

LOCALITY Near Mt Morgan, northern Queensland

These isolated plesiosaur vertebrae were thought to be of Upper Jurassic or Early Cretaceous age when first described by Bartholomai (1966b). They are now known to be Early Jurassic, and represent the equally oldest-known freshwater sauropterygians in the world, along with the Kolane specimens (see below). The bone is weathered away from the specimens, so they are studied from casts of the mould in the conglomeratic rock.

TECHNICAL DATA The vertebrae have deep transverse processes, and the centra are about 6 cm in diameter. The vertebrae were thought close to the Cretaceous English form *Leptocleidus*. In 1996 Dr Arthur Cruickshank examined the specimens and found them to be most similar to a new genus from the Jurassic Oxford Clay of England that he was working on, as both forms have solid rib bones (pachyostotic condition) which would have assisted them with ballast for deep diving or resting on the sea floor. Otherwise they could belong to a nothosaur (A. Cruickshank, pers.comm.).

GENUS INDETERMINATE

SPECIES Indeterminate

AGE Uppermost Early Jurassic

LOCALITY Kolane Station, south central Queensland (Westgrove Ironstone member of the Evergreen Formation)

Two occurrences of plesiosaurids, both described by Thulborn and Warren (1980), are known from the Early Jurassic sediments on Kolane Property, south central Queensland—the same location that *Siderops* was found in. One of the plesiosaurid specimens comprises the scant remains of an incomplete skeleton, including vertebrae and parts of limbs and their girdle bones. A second specimen is similarly from an incomplete skeleton, and includes recognisable parts of limbs and girdle, vertebrae and ribs. The bones come from long-necked plesiosaurs, and could be either elasmosaurid or cryptocleidid (Dr A. Cruickshank, pers. comm. 1996). Jurassic plesiosaurs outside of Australia are typically marine. The association of plesiosaurs with a labyrinthodont (*Siderops*) is indeed unusual.

TECHNICAL INFORMATION Some of the vertebrae show the rear margin of the neural spine is vertically grooved, an elasmosaurid feature. The cervical vertebrae also are of shape and proportions indicative of one of the long-necked plesiosaurians, rather than of a shorter-necked pliosaurid.

SUPERFAMILY PLESIOSAUROIDEA

FAMILY ELASMOSAURIDAE

These are long-necked plesiosaurians which have about 72 neck vertebrae.

GENUS INDETERMINATE

SPECIES Indeterminate

AGE Middle Jurassic

LOCALITY Bringo Cutting, near Geraldton, Western Australia (Colalura Sandstone)

Two small, isolated vertebrae from plesiosaurs are known from the railway cutting exposure near Bringo, about 20 km east of Geraldton in Western Australia. These vertebrae, some of which are well-preserved, have recently been described by Long and

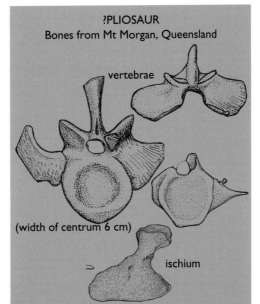

?PLIOSAUR
Bones from Mt Morgan, Queensland

vertebrae

(width of centrum 6 cm)

ischium

PLESIOSAURIAN CERVICAL VERTEBRA
From Bringo Cutting,
showing its main feature (width 6.3 cm)

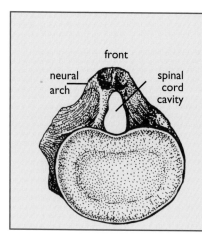

front

neural arch

spinal cord cavity

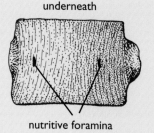

underneath

nutritive foramina

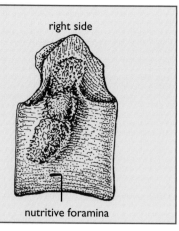

right side

nutritive foramina

Cruickshank (1998). The best specimen is a pectoral vertebra from an early elasmosaurid, the oldest recorded for Australia. Another specimen is a proximal caudal vertebra of an indeterminate plesiosaurian family. They occur in a coarse sandstone that was deposited in a shallow sea environment, probably near a large river mouth, since terrestrial dinosaurs are also known from the same deposit.

TECHNICAL DATA The very wide centrum is 63 mm across by 42 mm high in the midline. It shows a large broken area where the articulation surface is for the attachment of the pectoral rib, high up on the lateral faces of the centrum. The anterior zygapophysis is identified by the roughened area of bone on the anterior face of the neural arch, and the pectoral rib facet lies on both the neural arch and the centrum. It is very large, being equivalent in length to the midline height of the centrum. The wide proportions of the bone suggest it is most likely an elasmosaurid (Long and Cruickshank, 1998).

▲ Plesiosaurian cervical vertebra, possibly an early elasmosaurid, from Bringo Cutting, Western Australia (width 6.3 cm)

KRIS BRIMMELL

DINOSAURS

CLASS	REPTILIA
ORDER	SAURISCHIA
SUBORDER	SAUROPODOMORPHA

FAMILY ?CETIOSAURIDAE

The cetiosaurs ('whale lizards') are regarded as primitive sauropods. It is possible that the group may not form a natural family but could, instead, be just a 'mixed bag' of primitive and poorly known dinosaurs (Upchurch 1994).

GENUS RHOETOSAURUS

SPECIES *Rhoetosaurus brownei* Longman 1926

AGE Middle Jurassic

LOCALITY Eurombah Creek, Taloona Station, Roma District, Queensland (Injune Creek Beds)

The history of the discovery of *Rhoetosaurus* is briefly related in chapter 3. The fragmentary skeleton consisted of nearly complete leg and pelvic bones, about 28 consecutive tail vertebrae, 7 incomplete dorsal vertebrae and bits of rib, and one partial neck (cervical) vertebra. Enough of the beast was present for Heber Longman to name it as a new genus, *Rhoetosaurus*, after Rhoetos, one of the giants in Greek mythology, and 'sauros' (lizard). The species name honours Mr Browne for sending in the bones to the Queensland Museum. Since the early work by Longman, Dr Mary Wade of the Queensland Museum rediscovered the exact site in 1975 and found more bones from the original skeleton. Another excavation with Dr Anne Warren and Dr Zhao Xijin produced a second cervical vertebra. The new bones are currently under study by Dr Tony Thulborn.

Rhoetosaurus is Australia's most complete sauropod dinosaur, and had an estimated total length of about 12–15 m, with a hip height of about 4 m. An estimated weight of up to 20 tonnes would not be unlikely. The neck vertebra is quite elongate, suggesting the neck was probably long. It is also one of the world's earliest sauropods, and exhibits several primitive features in its skeleton. It is probably one of the primitive stock of sauropods—like *Cetiosaurus* from the Upper Jurassic of Oxfordshire in England—and does not show any specialised features which might relate it to any of the distinctive families of sauropods. Hunt *et al.* (1994) place it in the Cetiosauridae. Many of the bones of *Rhoetosaurus* possessed hollow cavities to reduce the overall weight of the skeleton, which, however, indicates that *Rhoetosaurus* was not as primitive as other cetiosaurs. The restoration of *Rhoetosaurus* below here is based on an idea of what a generalised sauropod is thought to have looked like, with its lack of some of the specialised features of the better-known sauropod families. Molnar (1991) comments that *Rhoetosaurus* 'was a very peculiar beast'. The short, stiff tail was suggested by Longman as possibly being used to prop the animal up, in the same manner as a kangaroo sometimes uses its tail. However, in the light of the fairly recent discoveries of club-tailed sauropods in China (for example, *Shunosaurus*), it would not be surprising if the short, stout tail of *Rhoetosaurus* also bore such a weapon. Only future discoveries might confirm such a speculative idea.

TECHNICAL DATA The tail rapidly tapers and was somewhat rigid. The genus is distinguished from all other sauropods by its caudal vertebrae having the following suite of characters. Anterior vertebrae are amphicoelous, with solid centra and with expanded elliptical articulating surfaces, the centra being somewhat compressed laterally. Prezygapophyses are elongated with vertical articulating surfaces, postzygapophyses absent, and the hyposphene is well-developed. Neural spines are robust, not elongated, and anterior ones are subrectangular in lateral view, with an oval median recess on the posterior margin above the junction with the hyposphene. Anterior chevrons are massive, not elongated, not confluent with their vertebral attachment. The neural canal is relatively

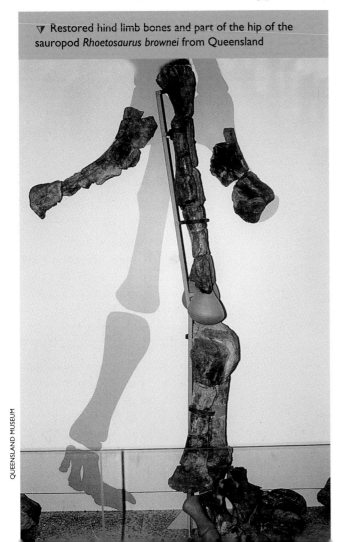

▽ Restored hind limb bones and part of the hip of the sauropod *Rhoetosaurus brownei* from Queensland

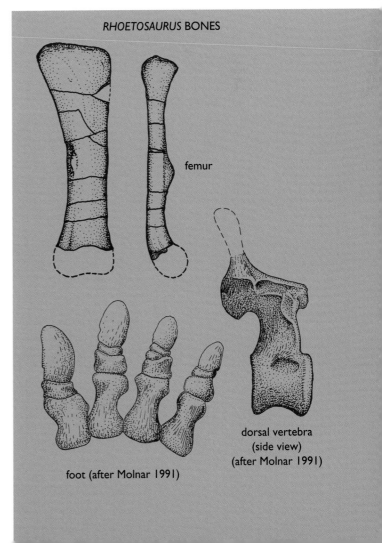

RHOETOSAURUS BONES

femur

foot (after Molnar 1991)

dorsal vertebra
(side view)
(after Molnar 1991)

large in anterior caudals. In addition, the dorsal vertebrae are opisthocoelous with lateral pleurocoels, with complex neural arches that have bracing laminae, small elevated zygapophysial articulations, and extensive intramural cavities.

GENUS INDETERMINATE

SPECIES Indeterminate

AGE Middle Jurassic

LOCALITY Bringo Cutting, near Geraldton, Western Australia (Colalura Sandstone)

This specimen is a small caudal vertebra probably, it is believed, from a sauropod (Long 1992). The specimen is 6.3 cm long by 4.7 cm wide. Although devoid of diagnostic features, it was compared with a number of other dinosaur caudal vertebrae and appeared most compatible with a similar distal caudal from the sauropod *Austrosaurus*, so it was identified as being possibly sauropod.

MIDDLE JURASSIC SAUROPOD?
Caudal vertebra from the Geraldton region, Western Australia

length 6.3cm

KRIS BRUMMELL

FAMILY INDETERMINATE

GENUS OZRAPTOR

SPECIES *Ozraptor subotaii* Long & Molnar 1998

AGE Middle Jurassic (Bajocian)

LOCALITY Bringo Cutting, near Geraldton, Western Australia (Colalura Sandstone)

This specimen was found in 1967 by Scotch College schoolboys. They handed it in to Prof. Rex Prider of the University of Western Australia, who had a cast made and sent it in to the British Museum (Natural History). The verdict was that the specimen was possibly that of a turtle bone. In recent years the author prepared it out of the rock and restudied it, identifying the specimen as that of a theropod dinosaur bone. It is a distal end of a shin bone (tibia) which shows the characteristic features of how the ankle bone (astragalus) attached to the leg. The name *Ozraptor* means 'Australian thief' (or colloquially dubbed 'The Lizard of Oz'), pertaining to the agile nature of the theropod, which was about 2 m long. The species name is after 'Subotai', the swift running thief and archer from the *Conan the Barbarian* films. The anatomical features of the shin-ankle connection are like no other theropod dinosaur and suggests that this animal may represent a unique lineage of Gondwana dinosaurs that was fairly advanced for its age. It is also the oldest bone of a theropod dinosaur from Australia.

TECHNICAL DATA The genus is distinguished by having a high astragalar groove which is almost square-shaped, and the fact that this groove has a distinct vertical ridge. The medial buttress on the tibia is very narrow. The width of the distal end of the tibia is 4 cm. Specimen number UWA 82469.

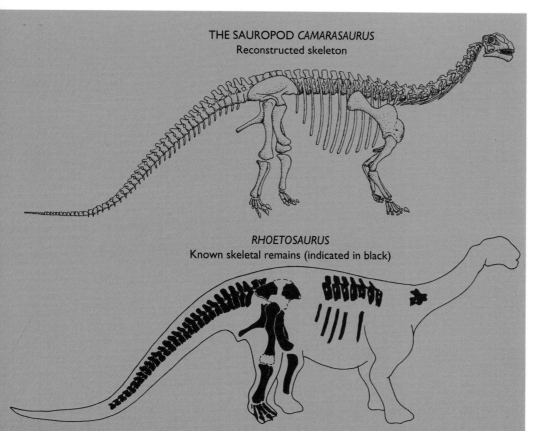

THE SAUROPOD *CAMARASAURUS*
Reconstructed skeleton

RHOETOSAURUS
Known skeletal remains (indicated in black)

Rhoetosaurus reconstructed
by Peter Schouten

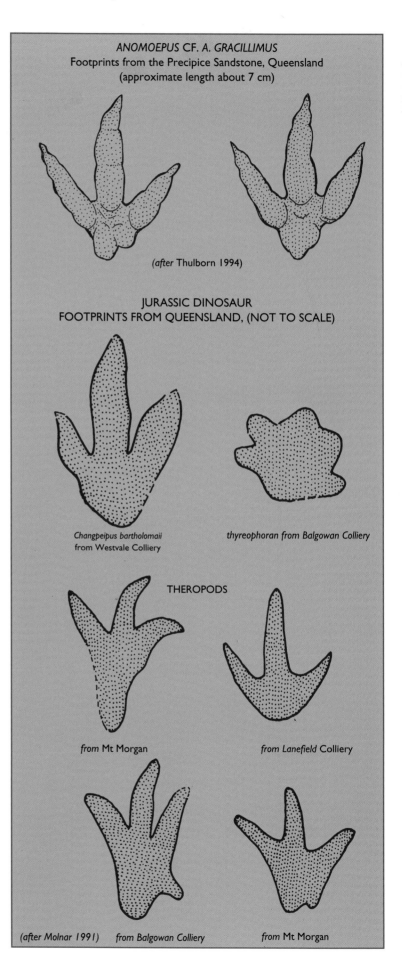

ANOMOEPUS CF. A. GRACILLIMUS
Footprints from the Precipice Sandstone, Queensland
(approximate length about 7 cm)

(after Thulborn 1994)

JURASSIC DINOSAUR
FOOTPRINTS FROM QUEENSLAND, (NOT TO SCALE)

Changpeipus bartholomaii
from Westvale Colliery

thyreophoran from Balgowan Colliery

THEROPODS

from Mt Morgan

from Lanefield Colliery

(after Molnar 1991)   from Balgowan Colliery   from Mt Morgan

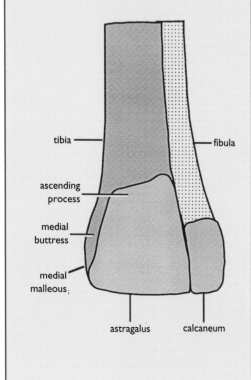

OZRAPTOR SUBOTAII
Middle Jurassic theropod tibia from the
Geraldton region, Western Australia
(width 4 cm)

tibia

fibula

ascending
process

medial
buttress

medial
malleous

astragalus

calcaneum

FOOTPRINTS

CLASS REPTILIA
ORDER SAURISCHIA
SUBORDER THEROPODA

FAMILY UNCERTAIN

FOOTPRINT GENUS CHANGPEIPUS

SPECIES *Changpeipus bartholomaii* Haubold 1971

AGE Middle Jurassic

LOCALITY Westvale No. 5 Colliery, southeastern Queensland (Walloon Group)

The name *Changpeipus* is not that of a dinosaur, but a form genus for a particular footprint type representing a large, meat-eating, three-toed dinosaur first described from China. The species name honours Dr Alan Bartholomai of the Queensland Museum.

TECHNICAL DATA Digit IV is longer and broader than digit II; digit III (middle) is much longer than other digits, and the distal ends of digits are relatively wide.

SUBORDER ORNITHOPODA

ICHNOFAMILY ANOMOEPODIDAE

FOOTPRINT GENUS ANOMOEPUS HITCHCOCK 1848

SPECIES *Anomoepus* cf. *A. gracillimus* Hitchcock 1844

AGE Lower Jurassic

LOCALITY Carnarvon Gorge (Carnarvon National Park), 95 km south of Rolleston, Queensland (Precipice Sandstone)

Tracks of *Anomoepus* were originally defined from rocks thought to be of Late Triassic age in the northeastern USA (but now thought to be Early Jurassic), and the genus has since been recognised in the Lower Jurassic of France and Southern Africa. The Queensland tracks comprise seven footprints on a slab, preserved as natural casts of fairly uniform size and shape. They were described in detail by Thulborn (1994), who believes the animals were about 30 cm high at the hip and less than 1.3 m long, probably small plant-eating dinosaurs

similar in appearance to *Fabrosaurus* of Southern Africa. The prints form parts of trackways and suggest walking speeds of about 2.5–2.9 km/h. The tracks are the oldest evidence of ornithopod dinosaurs from the Australian continent.

TECHNICAL DATA The prints vary in maximum length from 6.4 to 7.2 cm. The most complete prints show three toes with narrow claws, and a heel impression (called the metatarso-phalangeal node). Digits III and IV are weakly curved with a noted asymmetry of the digital nodes (two nodes in digit II, three in digit III and maybe four in digit IV). Narrow triangular claws are bluntly rounded, and the claw on digit III is medially deflected 30° from its main axis (see Thulborn 1994 for more details).

OTHER JURASSIC DINOSAUR FOOTPRINTS

Other meat-eating dinosaur prints from the Jurassic of Queensland come from Balgowan, Darling Downs region, and are up to 71 cm in length, indicating an animal about 12 m in length, making it one of the largest Jurassic predators known, on a par with *Allosaurus fragilis* from North America. The largest meat-eater from the Cretaceous Period, *Tyrannosaurus rex* (length 14–15 m), would have had footprints up to 80 cm in length, based on the actual size of its foot skeleton. Balgowan Colliery has also yielded small theropod tracks, some approximately 12 cm long, and a single, broad print from a quadrupedal dinosaur, with five digits preserved, possibly a primitive thyreophoran. Other theropod prints are recorded from the Lanefield Extended Colliery, south Queensland (Molnar 1991). The only other Jurassic dinosaur prints recorded outside of the Brisbane and Darling Downs area come from Mt Morgan, northern Queensland, where the handprints and footprints of a theropod are preserved. Originally they were thought to be Cretaceous in age, but are now dated as Early Jurassic. The handprint shows five digits present, indicating a primitive theropod.

part four

AUST
NEW ZE
in the CRET

RALIA and

ALAND

CEOUS

During the Cretaceous Period (144–65.4 million years ago) Australia began rifting away from Antarctica, but did not completely separate until just into the Tertiary Period. Sea-levels rose and formed a great inland sea running through the centre of Australia; central and southern Queensland was, at this time, a series of islands surrounded by this shallow inland sea. This sea deposited shallow marine sediments over much of Queensland, New South Wales and South Australia, the same sediments that are now a principal source of Australia's known dinosaurs—their dried carcasses presumably having been carried out to sea after flood events.

The climate was temperate at this time; although, as southeastern Australia was situated close to the Cretaceous South Pole, this region is seen as being temperate to cool (Rich *et al.* 1988). The plants were still predominantly conifers and pines and ferns, although the first flowering plants had appeared by the Early Cretaceous. Some of the world's oldest fossil flowers are known from the Koonwarra site, in eastern Victoria.

Cretaceous animal fossils are known throughout inland Queensland, and in the opal fields of New South Wales and South Australia. Large river systems developed in the rift valley as Australia and Antarctica began separating, and sediments carried by these rivers were deposited off the eastern and western coasts of Victoria. The Victorian dinosaurs have been found in these Early Cretaceous sediments of the Otway and Strzelecki Groups. Recently labyrinthodont amphibian remains have been recognised from these deposits, making them the youngest known occurrence of the group (see chapter 7).

In Western Australia Cretaceous reptiles are very rare: only a handful of dinosaur bones have been recovered from the Late Cretaceous Miria Formation and partially articulated plesiosaurs and ichthyosaurs, from the Early Cretaceous Birdrong Sandstone—both cropping out in the Carnarvon Basin.

Most of the known fossil vertebrates from New Zealand come from the Late Cretaceous, dated at around 65–72 million years old, a time poorly represented by fossils in Australia, and thus giving us a window into the southern world just before the dinosaurs, pterosaurs and the great marine reptiles became globally extinct. New Zealand

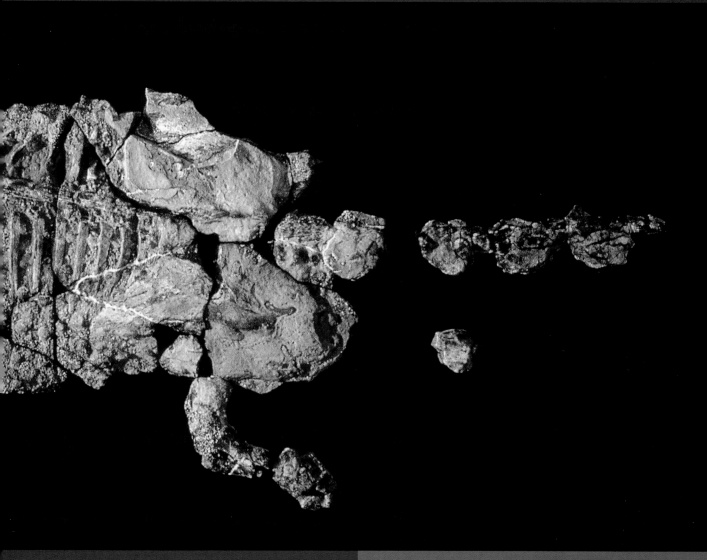

was then an island, which had just separated from western Antarctica, at about 82–85 million years ago. New Zealand was then situated within the Antarctic Circle, at latitude at least 66° south, and with estimated mean sea-water temperatures of 14°C, and with similar average land temperatures, which may have dropped to 10°C in winter, with occasional frosts and periods of colder spells (Molnar and Wiffen 1994). The land at this time would have been draped in *Nothofagus* and podocarp forests. By the close of the Cretaceous Period New Zealand had drifted considerably further north and was probably out of the Antarctic Circle.

▲ Newly found skeleton of *Minmi paravertebra* from near Richmond (approximate length about 2.5 metres)

▽ Australia during the Cretaceous Period, showing areas of land and sea, and locations of major animal fossil sites

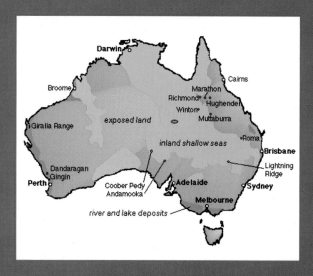

six

DINOSAURS

THEROPODS

ORDER SAURISCHIA
SUBORDER THEROPODA

FAMILY ALLOSAURIDAE

Allosaurids include many large meat-eaters of the Late Jurassic to Late Cretaceous. These are more advanced than primitive theropods in the following respects: they have a triangular 'boot' on the pubis that is longer anteriorly than posteriorly, and reduced forearms (with three digits on the hand), and a specialised foot with a reduced fifth digit).

GENUS ALLOSAURUS

SPECIES *Allosaurus* sp.

AGE Early Cretaceous (Aptian/Albian)

LOCALITY 100 m west of Eagles Nest, between Inverloch and Cape Paterson, eastern Victoria (Strzelecki Group)

Although as early as 1906 Woodward described a claw from a theropod dinosaur from Cape Paterson, Victoria, it was not until 1979 that a definite theropod bone of known affinities was found, discovered by Tim Flannery. The bone, which looked like an unidentifiable mess in the rock when it was found, turned out after preparation to be a nearly complete ankle bone (astragalus) of *Allosaurus* (Molnar *et al.*, 1981) Some controversy ensued over its identification (Welles 1983, Molnar *et al.* 1985). The theropod dinosaurs have a distinctive ankle bone, which has a high ascending process on the front face that covers the lower end of the shin bone (tibia), and this bone is distinctive for each of the different genera.

THEROPOD DINOSAUR FOOT
Position of the *astragalus*

VICTORIAN *ALLOSAURUS*
astragalus

front

back

The Victorian *Allosaurus* was more robust than the North American species, but somewhat smaller—at about 5–6 m maximum length, and about 2 m high, compared to a length of 12 m for the North American species *A. fragilis*. The *Allosaurus* species known from North America died out at the end of the Jurassic Period, making the Victorian *Allosaurus* another example of 'a living fossil' which survived longer in Australia than elsewhere. As the Victorian dinosaurs lived in close proximity to the Cretaceous South Pole (Rich *et al.* 1988), it is quite feasible that they may have had a fur or feather-down covering as an adaptation for cold weather (as depicted in Long 1993). Other archosaurs had the ability to develop 'fur' (for instance, in some pterosaurs), and recently a small feather-covered theropod (*Sinosauropteryx prima*) has been discovered from China; the growth of such coats on certain dinosaurs is not, then, as unlikely as it might at first appear. In recent years, no additional material of the Victorian *Allosaurus* has been found, although part of a very large pedal claw was recently found. The top of the claw was sheared away by erosion, but enough of it remains to indicate that it would have been about 15 cm in length, indicating a much larger theropod also existed alongside the allosaurid.

▲ Posterior view of astragalus (ankle bone) of *Allosaurus* sp. from Victoria (width 10 cm)

DR TOM RICH AND DR PAT VICKERS-RICH

TECHNICAL DATA The astragalus of members of the family Allosauridae is distinguished from those of all other theropods by the restriction of the ascending process to the lateral part of the bone; the medial condyle is large with respect to the lateral condyle, and there is a lower horizontal groove across the face of the condyles (Molnar *et al.* 1981). The Victorian astragalus (Museum of Victoria P150070) more closely resembles that of *Allosaurus fragilis* than any other form in six main features: fibular facet distinct; high, but medially restricted, ascending process; ascending process has an inflection on its medial margin; medial condyle much larger than lateral one; calcaneal notch well-defined, and distinct groove across the condyles. Molnar *et al.* (1981) also point out six differences between this specimen and the astragalus of *A. fragilis* and conclude that, although these reflect differences in the ligamentous mode of attachment between the astragalus and tibia, the mode of interlocking between the calcaneum and astragalus is a specialised condition seen only in the two forms, and should be sufficient evidence that the Victorian specimen belongs in the genus *Allosaurus*.

FAMILY ?COELURIDAE

Coelurosaurs were small, lightly built carnivorous dinosaurs which could run fast. They were often thought to be primitive amongst theropods, but are now classified as being more specialised than the allosaurids, megalosaurids, abelisaurids and ceratosaurs. The group Coelurosauria includes *Compsognathus*, dromaeosaurids, oviraptorids, ornithomimosaurids, troodontids, elmisaurids and tyrannosaurids (Holtz 1994). The skull has large orbits for the eyes and is generally slender, and the hands and feet are often less specialised than in later theropods.

GENUS RAPATOR

SPECIES *Rapator ornitholestoides* von Huene 1932

AGE Early Cretaceous

LOCALITY Lightning Ridge, New South Wales (Griman Creek Formation)

A single hand bone, preserved in opal, was named and described by a German dinosaur expert, Friederich von Huene, in 1932. It had such a striking resemblance to that of the coelurosaur *Ornitholestes* from North America that it received the species name meaning 'like *Ornitholestes*'. *Rapator* is probably derived, in error, from the Latin 'raptor', which means 'thief' or 'plunderer'. In size *Rapator* would have been similar to a small adult *Allosaurus*, about 9 m in length. Other theropod bone fragments from Lightning Ridge could most likely belong with *Rapator* according to Molnar (1982a, 1991). The relationships of this dinosaur are unclear at present, although it appears to be completely different from other families known at this time (e.g., allosaurids, carcharodontosaurids).

TECHNICAL DATA *Rapator* is characterised by its first metacarpal bone having an elongated posteromedial process, a feature otherwise seen only in *Ornitholestes*. It differs from that genus, however, in that the first metacarpal bone in *Rapator* is considerably larger, broader relative to length, more robust, and with the posteromedial process more prominently developed.

GENUS KAKURU

SPECIES *Kakuru kujani* Molnar and Pledge 1980

AGE Early Cretaceous

LOCALITY Opal Fields of Andamooka, South Australia (Maree Formation)

RAPATOR ONITHOLESTOIDES, TYPE SPECIMEN
Metacarpal bone (from the hand)

length of bone 7 cm

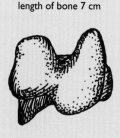

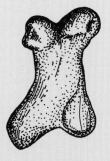

| postero medial view | posterior end | anterior end | dorsal view | ventral view |

The beautiful, opalised leg bone of *Kakuru* was first seen by a scientist when Neville Pledge, of the South Australian Museum, spied the specimen in an Adelaide opal shop. Luckily, he obtained permission from the owner, Mr A. Fleming, to make a cast of the bone, for the specimen was soon sold at auction and disappeared without trace. The name *Kakuru* is from a Guyani Aboriginal word meaning 'rainbow serpent', and the species name is a translation of the name 'Guyani'.

Kakuru is based on a nearly complete shin bone (tibia) with slender proportions—similar to the bones from modern wading birds—and, as the shape of this region is unique to each genus of theropod, it is clear that *Kakuru* is a distinct and valid genus of dinosaur. In size, *Kakuru* was small, probably only 2–3 m in length. In overall shape and proportions the tibia of *Kakuru* closely resembles that of the bird-like dinosaur '*Avimimus*' from the Late Cretaceous of Mongolia (although this genus may not be valid according to Mark Norell, American Museum of Natural History, pers. comm. 1996). However, Molnar and Pledge (1980) state that if the structure of the tibia has been correctly interpreted, having such a high ascending process on the astragalus, then *Kakuru* does not readily fit into any of the five classes of theropod ankle types recognised by Welles and Long (1974); it may therefore belong to a completely new lineage of theropod. A small toe bone (phalanx) and a fragment possibly belonging to a fibula are also tentatively included with *Kakuru* as they came from the same site and have similarly slender proportions.

TECHNICAL DATA The type specimen is now a cast of the original held by the South Australian Museum (no. P17926, called a 'plastoholotype'). The distal end of the tibia measures about 44 mm wide, and has a distinctive facet for the astragalus,

which is high and attenuated to a distinct dorsal apex, but not broad enough to extend across the entire width of the anterior distal end of the tibia; it is medially limited by a marked anterior ridge of the tibia, which runs dorsally from the strongly produced medial malleolus.

Reconstruction of a slender coelurosaur, indicating an animal comparable to *Kakuru*

GENUS INDETERMINATE

SPECIES Indeterminate

AGE Early Cretaceous

LOCALITY Opal Fields of Coober Pedy, South Australia (Maree Formation)

This specimen is part of the end of a foot bone (metatarsus) and closely resembles that of a theropod. In overall size (breadth 39 mm), and in its slender proportions, it could well belong with *Kakuru*, which comes from rocks of the same age in nearby Andamooka. I have examined a cast of the bone kindly sent to me by Neville Pledge, of the South Australian Museum, but have no information about its discovery.

TECHNICAL DATA The specimen is registered as South Australian Museum no. P35321. It appears to be the distal end of the fourth metatarsus, and compares closely with forms like *Avisaurus* from the Late Cretaceous of North America in having an expanded distal end to the bone, with extreme thinness of the shaft (as defined by Norman 1990).

NEXT PAGE *Allosaurus* stalking in the cool polar forests of southern Victoria 100 million years ago. By Tony Windberg

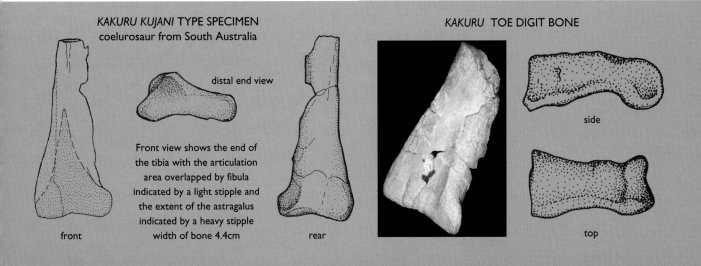

KAKURU KUJANI TYPE SPECIMEN
coelurosaur from South Australia

distal end view

Front view shows the end of the tibia with the articulation area overlapped by fibula indicated by a light stipple and the extent of the astragalus indicated by a heavy stipple
front   width of bone 4.4cm   rear

KAKURU TOE DIGIT BONE

side

top

A. Windbee
© 98

FAMILY ORNITHOMIMOSAURIDAE

Ornithomimosaurs were the long-necked, long-legged, ostrich-like dinosaurs ('ornithomimo' means 'bird mimic'). They mostly lacked teeth, and probably had a horny bill over the jaws. The eyes were very large. They are thought to be the fastest-running of all dinosaurs, with estimated speeds of 60–100 km/h, based on trackway data, although the top end speed estimates (by Robert Bakker) are disputed by some palaeontologists. They probably fed on insects, eggs, and small animals. Most species come from the Late Cretaceous of North America and Mongolia.

GENUS TIMIMUS

SPECIES *Timimus hermani* Rich and Vickers-Rich 1994

AGE Early Cretaceous (Albian)

LOCALITY Otway Ranges, coastal exposure at Dinosaur Cove, Victoria (Otway Group)

Timimus was named in honour of Timothy Rich and Tim Flannery, and from the Greek 'mimos', meaning 'mimic', with the species name honouring John Herman. All other ornithomimosaurs are from the Late Cretaceous of North America and Asia, except possibly for *Elaphrosaurus* from

Timimus femora (holotype, Museum of Victoria no. P186303) are the absence of an intercondylar groove on the anterior (extensor) surface, suggesting that it is more primitive than all other ornithomimosaurs. Bone growth studies of *Timimus* indicate that periosteal growth was discontinuous and apparently cyclical, with traces of fast initial growth (Chinsamy *et al.* 1996).

FAMILY OVIRAPTOROSAURIDAE

Oviraptorosaurs were strange-looking theropods known principally from the Late Cretaceous of Asia and North America. They had unusual lower jaws with long angular and surangular bones and short, toothless dentaries. It has been suggested that they were egg-eaters, as one early discovery in Mongolia was of an oviraptorid skeleton near fossilised dinosaur eggs thought to belong to another dinosaur, *Protoceratops*. Recent finds by Mark Norell and his team from the American Museum of Natural History in the Gobi Desert, Mongolia, indicate that the associated oviraptorid skeleton with nests of eggs probably represent animals brooding their own nests (Norell *et al.* 1995). The discovery of possible oviraptorosaurs from Victoria is exciting since it raises the prospect

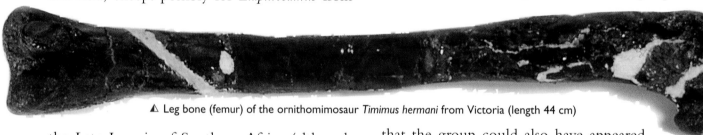

▲ Leg bone (femur) of the ornithomimosaur *Timimus hermani* from Victoria (length 44 cm)

the Late Jurassic of Southern Africa (although Holtz [1994] has recently cast doubt on whether this genus has any relationship to ornithomimosaurs). Thus, the finding of *Timimus* was of great interest in the scientific world, as it led to the suggestion that the ornithomimosaur group may have first originated in Gondwana before migrating northwards to Asia and North America.

The genus is based upon a complete left femur about 44 cm in length (comparable in size with a human femur), indicating an animal close to 2.5 m in length, perhaps standing around 1.7 m high. A second, smaller femur, about 19.5 cm long, was found less than one metre away from this specimen. Vertebrae of possible ornithomimosaurs are also known from the Dinosaur Cove sites, and most likely belong to *Timimus* also.

TECHNICAL DATA The distinctive features of the

that the group could also have appeared here earlier than in the Northern Hemisphere. More distinctive remains are needed, however, to confirm the finds.

GENUS INDETERMINATE

SPECIES Indeterminate

AGE Early Cretaceous (Albian)

LOCALITY Otway Ranges, coastal exposure at Dinosaur Cove East locality, western Victoria (Otway Group)

Part of a lower jaw bone of a possible oviraptorosaur from Victoria was recently described by Currie *et al.* (1996). If complete, the entire lower jaw would have measured about 20 cm, suggestive of an animal about 2 m long. The jaw shows intermediate features between dromaeosaurids and oviraptorosaurs. In addition to this enigmatic jaw fragment, an isolated vertebra of a theropod from the same locality has been tentatively identified as belonging to the same group.

TECHNICAL DATA The jaw fragment is part of a right surangular bone. Although this bone is incomplete anteriorly and posteriorly, with its curvature accentuated by slight distortion, the specimen still resembles the unusual surangular seen in oviraptorosaurs, in having a coronoid process with a medial inflection. Its slender form is more reminiscent of the lower jaw in caenagnathids rather than advanced oviraptorids like *Oviraptor* or *Ingenia*; yet it also differs from all known oviraptorosaurs in not being fused to the articular and coronoid bones. The dorsal vertebra has simple internal cavities, and possesses a low, poorly developed lamina between the parapophysis and the posteroventral corner of the neural arch, a feature not seen on any dromaeosaur but one that is present on *Oviraptor philoceratops*.

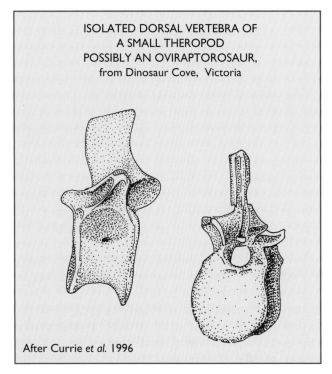

ISOLATED DORSAL VERTEBRA OF
A SMALL THEROPOD
POSSIBLY AN OVIRAPTOROSAUR,
from Dinosaur Cove, Victoria

After Currie *et al.* 1996

▲ Reconstruction by Peter Trusler of a generalised ornithomimosaur, showing how *Timimus* may have looked

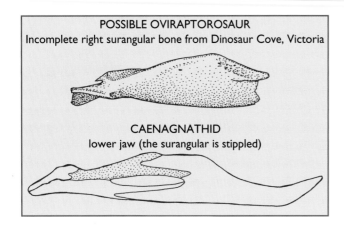

POSSIBLE OVIRAPTOROSAUR
Incomplete right surangular bone from Dinosaur Cove, Victoria

CAENAGNATHID
lower jaw (the surangular is stippled)

FAMILY DROMAEOSAURIDAE

Dromaeosaurids were small to medium-sized theropods with well-developed sickle claws on the inner toe of the foot, as well as particularly large claws on the hands. This group, which Holtz (1994) considers is the most closely related dinosaur group to birds, includes the well-known genera *Velociraptor* and *Deinonychus*.

GENUS INDETERMINATE

SPECIES Indeterminate

AGE Early Cretaceous (Aptian/Albian)

LOCALITY Otway Ranges, coastal exposure at Dinosaur Cove East locality, western Victoria (Otway Group)

Dr Phil Currie working with Dr Tom Rich and Dr Pat Vickers-Rich have identified isolated theropod teeth as belonging to small dromaeosaurids, based on the nature of their serration patterns and curvature. Some other isolated bones of theropods from the Dinosaur Cove site could belong with these teeth, but association between the two is impossible to demonstrate from the isolated nature of the individual finds.

TECHNICAL DATA The teeth are moderately recurved with coarse serrae on the posterior cutting edge of each tooth, but are not serrated anteriorly.

FAMILY INDETERMINATE

GENUS WALGETTOSUCHUS

SPECIES *Walgettosuchus woodwardi* von Huene 1932

AGE Early Cretaceous

LOCALITY Lightning Ridge, New South Wales (Griman Creek Formation)

The name *Walgettosuchus* was given by von Huene to a poorly preserved tail vertebra from Lightning Ridge, which, unfortunately, displays no diagnostic features that could characterise a new genus. Molnar (1982a) comments that it could conceivably belong to one of three families of theropod (Allosauridae, Coeluridae, or Ornithomimidae), so its status as a separate genus is highly doubtful. However, as additional theropod material becomes available from Lightning Ridge perhaps the taxonomic positions of *Walgettosuchus* and *Rapator* might one day be resolved. The name comes from Walgett, in northern New South Wales and 'suchus' meaning 'crocodile'.

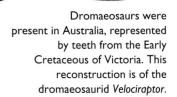

Dromaeosaurs were present in Australia, represented by teeth from the Early Cretaceous of Victoria. This reconstruction is of the dromaeosaurid *Velociraptor*.

GENUS INDETERMINATE

SPECIES Indeterminate

AGE Early Cretaceous (Hauterivian–Barremian)

LOCALITY North of Kalbarri, Western Australia (Birdrong Sandstone)

An incomplete, small tail vertebra of a possible theropod dinosaur was reported from the Early Cretaceous of Western Australia by Long and Cruickshank (1996), but it not known to which

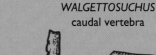
◁ Type specimen of *Walgettosuchus woodwardi*, an indeterminate vertebra

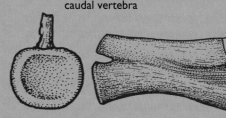

WALGETTOSUCHUS
caudal vertebra

JOHN A. LONG

▲ Indeterminate theropod fourth toe bone from the Late Cretaceous, near Gingin, Western Australia (length 4 cm)

theropod family it might belong. The specimen is the only Early Cretaceous dinosaur bone from Western Australia, and was found in shallow marine sandstones containing the pliosaurid *Leptocleidus clemai* and ichthyosaur *Platypterygius* sp.

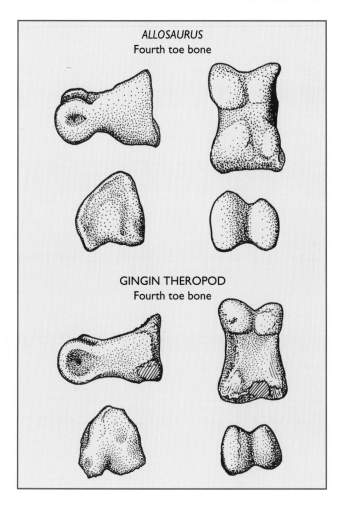

GENUS INDETERMINATE

SPECIES Indeterminate

AGE Late Cretaceous (?Cenomanian–Santonian)

LOCALITY Molecap Quarry, near Gingin, Western Australia (Molecap Greensand)

A single toe bone from a theropod dinosaur was found in the Molecap Hill Quarry near Gingin, 90 km north of Perth, in 1992 by Michael Green, a University of Western Australia geology student. The bone was found protruding from the greensand about half-way up the west side of the quarry wall. It is the first dinosaur bone to be found in the Perth Basin.

TECHNICAL DATA The bone measures 40.8 mm in length and is well-preserved (Long 1995b). It is the fourth pedal proximal phalanx from the left foot of a moderate-sized theropod which is closely comparable with *Allosaurus,* but much smaller (about 4 m).

GENUS INDETERMINATE

SPECIES Indeterminate

AGE Late Cretaceous (Late Maastrichtian)

LOCALITY Giralia Range, south of Exmouth Gulf, Western Australia (Miria Formation)

A weathered half of a dinosaur upper arm bone (humerus), measuring 21 cm long, was found in the Giralia Range by Mr George Kendrick in August 1990, and described by Long (1992). It is significant in being the youngest known dinosaur bone from Australia, and the only theropod humerus known from the country. It probably came from either a medium-sized dromaeosaur with large arms, or a larger tetanuran with small arms.

TECHNICAL DATA The specimen represents the proximal part of the humerus, showing the outline of the prominent deltopectoral crest, and has a slender, rapidly tapering shaft.

▽ Indeterminate theropod humerus from the Late Cretaceous, Giralia Range, Western Australia. Estimated bone length is 35 cm (restored)

SAUROPODS

SUBORDER SAUROPODOMORPHA

FAMILY ?TITANOSAURIDAE

Titanosaurids appear to constitute the only sauropod family represented in the Late Cretaceous, although recently an endemic group from this age in Mongolia was referred to the new family Nemegtosauridae (Upchurch 1994). Titanosaurids have extremely robust front limb bones and some Gondwanan forms reached gargantuan sizes, placing them amongst the largest land animals that ever lived (for example, *Argentinosaurus*). Coombs and Molnar (1981) tentatively placed the specimens from Winton—referred to as *Austrosaurus* sp.—in the subfamily Cetiosaurinae, although it seems more probable that they are titanosaurids (Hunt *et al.* 1994).

GENUS AUSTROSAURUS

SPECIES *Austrosaurus mackillopi* Longman 1933

AGE Early Cretaceous

LOCALITY On 'Clutha' Station, approximately 60 km from Maxwelton, west of Townsville, northern Queensland (Allaru Mudstone; probably also Winton Formation)

AUSTROSAURUS MACKILLOPI
back vertebra viewed from the front

The history of the discovery of *Austrosaurus* has already been given in chapter 3. The original remains of *Austrosaurus* comprise three massive blocks of rock, each containing paired dorsal vertebrae and some rib fragments. Molnar (1982a) comments that the bones of *Austrosaurus*, although they are from the top of the Early Cretaceous, resemble those of Middle Jurassic forms from overseas, as the centra are composed of spongy bone and the pleurocoels (hollows in the bone to reduce weight) are only rudimentary compared with contemporary sauropods.

Additional specimens found near Winton can probably be assigned to *Austrosaurus* sp. (Molnar 1982a), and consist of a wrist/ hand bone (metacarpus) and simple tail vertebrae. Coombs and Molnar (1981) described several isolated sauropod bones, including limbs, hip, shoulder girdle and vertebrae, from the early Upper Cretaceous Winton Formation, which they provisionally refer to *Austrosaurus* sp. The long hand bones suggest an animal with a long forelimb like the gigantic *Brachiosaurus*, a well-known genus from North America and Africa. Both brachiosaurids and cetiosaurids lack the robust forearm bone (radius) seen in the Winton sauropod, so the titanosaurids are the most appropriate family at present in which to place the collective material of *Austrosaurus*.

TECHNICAL DATA Longman's (1933) diagnosis of the type material is as follows: dorsal vertebrae markedly opisthocoelous; centra with thin cortical walls, much enlarged at the neural arch articulations; intramural region with a complex of small cavities; pleurocoels prominent, with external and internal divisions; neural arch with deep recess between the prezygapophysial lamina and the infradiapophysial buttress. In addition, Molnar (1991) notes that the tail vertebrae are simple in having no struts or buttresses supporting them, and lack pleurocoels. The metacarpal bones of the hand are also very large relative to the arm size and compared with other sauropods.

FAMILY BRACHIOSAURIDAE

Brachiosaurs were very large sauropods with front limbs larger than the hind limbs, and vertebrae that had hollowed-out centres to reduce weight. Brachiosaurids were amongst the largest known dinosaurs.

QUEENSLAND MUSEUM

Reconstructed front limb bones of *Austrosaurus* from Queensland

Left column side panel:

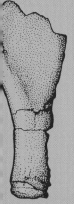

STROSAURUS sp.
arge sauropod
ones from near
ton, Queensland

erus (90 cm long)

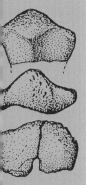

limb bone ends

radius

femur

GENUS INDETERMINATE

SPECIES Indeterminate

AGE Early Cretaceous

LOCALITY Near Hughenden, northern
Queensland (formation unknown)

The rear half of a large neck vertebra from near Hughenden closely resembles that of a *Brachiosaurus*, and suggests an overall size for the animal of close to 20 m long, which would make it Australia's largest known dinosaur. As Molnar (1982a) states, it is not possible to determine whether this bone actually belongs with *Austrosaurus,* as no neck bones of *Austrosaurus* are known. The resemblance to *Austrosaurus* and brachiosaurids is not as close as it is between the isolated Hughenden vertebra and brachiosaurids, so it is most likely that the Hughenden specimen constitutes an entirely different sauropod.

TECHNICAL DATA   Brachiosaurid cervical vertebrae typically lacked the forked neural spines for the neck ligament that occur in other sauropods. They have enormous pleurocoels, with very thin intramural walls, and the posterior face of the centrum may jut out well past the posterior limit of the neural spine. These features are seen on the Hughenden specimen (Queensland Museum specimen no. F6142).

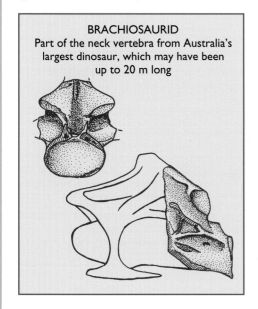

BRACHIOSAURID
Part of the neck vertebra from Australia's largest dinosaur, which may have been up to 20 m long

---

THYREOPHORANS

SUBORDER ANKYLOSAURIA

FAMILY ANKYLOSAURIDAE

Thyreophorans are the group of dinosaurs containing the armoured stegosaurs, the nodosaurs and ankylosaurs, and the ceratopsians. Ankylosaurids were small to medium-sized quadrupedal dinosaurs with armour plates in their skin, some with well-developed spines and the club tails. They had broad heads and grew to massive sizes, some—for instance, *Ankylosaurus* —being 10 m long.

GENUS MINMI

SPECIES *Minmi paravertebra* Molnar 1980a

AGE Early Cretaceous

LOCALITY The type specimen is from just south of Mack Gulley, north of Roma, central southern Queensland (*Minmi* member, Bungil Formation, Molnar 1980a); a nearly complete skeleton of a different species (*Minmi* sp., Molnar 1996b) was found near Richmond, central Queensland (Allaru Mudstone; also from Toolebuc Formation). In addition another five smaller specimens are also assigned to *Minmi*—Queensland Museum F33565, F33566 (maybe from the one individual, Molnar 1996b), Australian Museum F35259, Queensland Museum F33286, and one unregistered specimen.

The original specimen of *Minmi* was discovered by Dr Alan Bartholomai, of the Queensland Museum, in 1964 from a gulley near Roma. The name *Minmi* derives from Minmi Crossing, near this site. The skeleton was partially articulated, preserved in limey concretions, and consists of eleven vertebrae from the back with some associated ribs, and an incomplete right foot, and much of the ventral body armour. In November 1989 an almost complete skeleton of *Minmi* was found east of Richmond on Marathon Station, in limey shales, by Mr Ian Ievers, and was collected in January 1990. It is one of the world's most complete ankylosaurs, and includes the well-preserved skull and much of the body dermal armour. This specimen has been acid-prepared by the Queensland Museum. The total length *Minmi* would have been no more than about 2.5–3.5 m, and it is thought to represent possibly a newly mature individual. It may have been a

NATIONAL PHILATELIC COLLECTION, AUSTRALIA POST

▲ Reconstruction of *Minmi* with hypsilophodontids in the background. By Peter Trusler

mummified carcass, carried by flood waters after heavy rains, eventually reaching the sea, where it became waterlogged and sank.

The vertebrae of *Minmi* are unique amongst dinosaurs in having, alongside the neural spines, bony elements (called paravertebrae), which supported the armour plates on the creature's back (Molnar and Frey 1987). The belly was covered with numerous small ossicles or scutes about 5 mm in diameter, and the back bore different-sized scutes—particularly large ones are found on the neck (13 *x* 9 cm), over the shoulder, over the hip and at the base of the tail. Rows of medium-sized scutes occur over most of the body, with polygonal masses of small scutes in between these. The skull is one of the best of any ankylosaurs. It is almost pentagonal in dorsal view.

Despite its incompleteness *Minmi* is still the most complete ankylosaur known from the Southern Hemisphere, and indicates that ankylosaurid dinosaurs were widespread around the world soon after their first appearance in the fossil record (Neocomian, Early Cretaceous). There are possibly two species of *Minmi* present in the Queensland faunas—the one from the type locality, and the other occurring in the Toolebuc Formation and Allaru Mudstone (Dr Ralph Molnar, pers comm. 1996), which Molnar suggests is in essence probably an ankylosaur. As most of the known members of that group come from the last 10 million years of the Cretaceous in Asia and North America, an early form from Gondwana would be noticeably different, perhaps even representing an entirely new group of armoured dinosaurs.

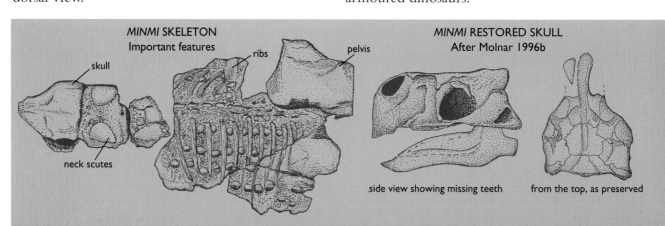

MINMI SKELETON
Important features

skull

neck scutes

ribs

pelvis

MINMI RESTORED SKULL
After Molnar 1996b

side view showing missing teeth

from the top, as preserved

A CERATOPSIAN?

SUBORDER CERATOPSIA

These include the well-known, large, horned dinosaurs, such as *Triceratops*, although the earliest members of the group were small and lacked horns. They evolved from the parrot-like psittacosaurids. Although most ceratopsians come from the Late Cretaceous of Asia and North America, the discovery of one in Australia of Early Cretaceous age raises the possibility of the place of origin of the group being from Gondwana (Rich & Vickers-Rich 1994).

SUPERFAMILY NEOCERATOPSIA

FAMILY ?PROTOCERATOPSIDAE

Protoceratopsids include those primitive ceratopsian dinosaurs lacking well-developed frills and horns. They were small plant-eating animals, generally under 2 m in length, and have been called the 'sheep' of the dinosaur world because their skeletons were so abundant in certain Cretaceous formations in Mongolia.

GENUS ?LEPTOCERATOPS

SPECIES Indeterminate

AGE Early Cretaceous (Aptian)

LOCALITY The Arch, coastal exposures near Inverloch, southeastern Victoria (Strzelecki Group)

A single, almost-complete ulna of a neoceratopsian dinosaur was described from Victoria by Rich and Vickers-Rich (1994). When the bone was first discovered, it remained a mystery until its finders took it across the world to the USA, where it was shown to various leading dinosaur experts and compared with Northern Hemisphere dinosaur skeletons. Dale Russell, a well-known North American dinosaur expert was taken aback when he saw the Australian bone lying next to the ulna of a Canadian ceratopsian. He exclaimed 'we are in violent agreement, that

TECHNICAL DATA *Minmi* is characterised by its vertebrae, in that they have paravertebral elements present; dorsal vertebrae are amphiplatan without notochordal knobs, transverse processes are slender and triangular (not T-shaped), the neural canal is broad, and the posterior intervertebral notch is shallow. The teeth have seven to nine well-pronounced denticles, with furrows on one side and are mostly smooth on the other side. The ventral armour was formed by a pavement of small ossicles. *Minmi* shares some ankylosaur features (for example, the snout arches higher than the skull), yet retains many primitive features expected to be found in an ancestral thyreophoran. Some of the primitive features seen in *Minmi* are: the femur is rounded in section; the skull is longer than wide; the acetabular and postacetabular regions of the ilium are both long (Molnar 1994).

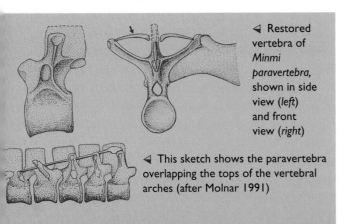

◁ Restored vertebra of *Minmi paravertebra,* shown in side view (*left*) and front view (*right*)

◁ This sketch shows the paravertebra overlapping the tops of the vertebral arches (after Molnar 1991)

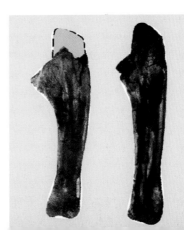

◁ Partial ulna of a neoceratopsian dinosaur from eastern Victoria (*left*) and the ulna of *Leptocercatops gracilis* from Canada (*right*)

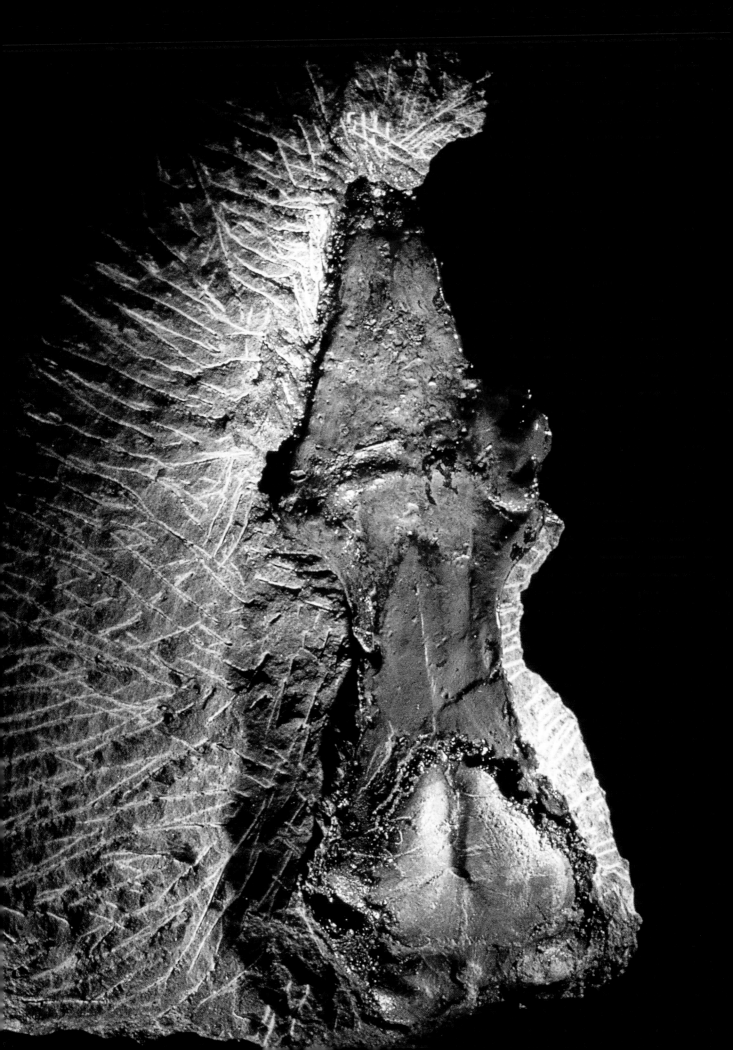

sure looks like a ceratopsian!'—a group thought to be unique to the Late Cretaceous deposits in North America and Asia. The bone is similar to that of *Leptoceratops gracilis* from Canada, an animal about 1.5 m in length.

To date no other ceratopsian bones have been found in Australian dinosaur sites. Alternatively, it may well be that this bone represents a parallel group of dinosaurs that evolved in Gondwana occupying the niche ceratopsians had elsewhere—similarly to the way that the marsupial Tasmanian tigers (thylacines) and dogs evolved independently the same body and head shape, without being even closely related. Until more bones of this enigmatic dinosaur are found, it is best left classified as 'most likely a neoceratopsian'.

TECHNICAL DATA The ulna, approximately 16 cm long as preserved (missing the olecranon), is short and deep, being mediolaterally flattened, a characteristic of neoceratopsians.

ORNITHOPODS

| ORDER ORNITHISCHIA |
| SUBORDER ORNITHOPODA |

FAMILY HYPSILOPHODONTIDAE

Hypsilophodonts were the fleet-footed gazelles of the dinosaur world. They were mostly small, plant-eating dinosaurs. Their teeth are characterised by being high-crowned with numerous ridges for chewing tough plant material.

GENUS LEAELLYNASAURA

SPECIES *Leaellynasaura amicagraphica* Rich & Rich 1989

AGE Early Cretaceous (Aptian/Albian)

LOCALITY Otway and Strzelecki ranges, coastal exposures (including Dinosaur Cove), Victoria (Otway and Strzelecki Groups)

◹ Articulated leg bones of a small hypsilophodontid from Victoria, possibly *Atlascopcosaurus*. (approximate length of tibia 20 cm)
◁ Skull of the Victorian hypsilophodont *Leaellynasaura amicagraphica*, viewed from above. (skull length just under 6 cm)

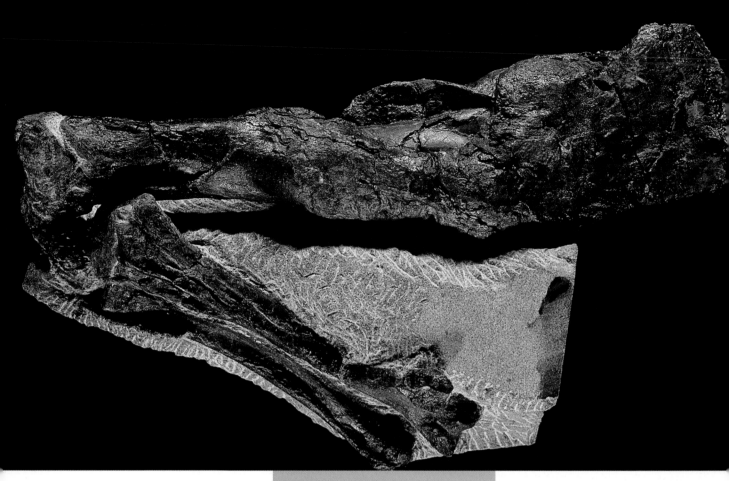

Leaellynasaura takes its name from Leaellyn Rich, who helped find some of the bones of this little hypsilophodont, which would have only been about a metre in length. The species name jointly honours the National Geographic Society, who funded much of the fieldwork at Dinosaur Cove, and the friends of the Museum of Victoria. The skeleton is known from many isolated limb bones, ribs, vertebrae, jaws, teeth and one partial skull. The skull of a juvenile individual is preserved in dorsal view, showing the large orbital area for the eyes, and the endocast of the brain cavity reveals an enlarged optic lobe, possibly an adaptation for coping with three months of the year in darkness. Because *Leaellynasaura* had to endure cold temperatures, the artist and I have reconstructed the animal with some feather-down covering, based on the similar development of feathers in small theropods like the newly discovered Chinese genus *Sinosauropteryx*.

▲ Articulated leg of a hypsilophodont dinosaur from Dinosaur Cove, Victoria, showing bone infection (osteomyelitis) of the femur (length of femur under 19 cm)

▼ Exacavating dinosaur bones at Dinosaur Cove, Victoria

STEVE MORTON/DR TOM RICH AND DR PAT VICKERS-RICH

Studies of the bone growth in the Victorian hypsilophodontids (Chinsamy *et al.* 1996) indicate that they had a continuous rate of periosteal bone growth, which then slowed after a burst of initial rapid growth. Comparisons with other hypsilophodontids from varying latitudes in the Cretaceous indicate that the Victorian forms, which had azonal fibro-lamellar bone growth instead of cyclical bone formation, may have had physiological specialisations to cope with changing environmental conditions. Some of the bones have shown cases of bone infection, or osteomyelitis (Gross *et al.* 1993); this might indicate that there was low density of animal populations around at the time since predators, if they were abundant, would certainly have made a meal out of a wounded hypsilophodont. The little fleet-footed hypsilophodonts from Victoria had to watch out for the meat-eater *Allosaurus* as well as smaller predatory dromaeosaurids, which inhabited the same wide, river valley environment.

ridges on them but lack central carinae. Dorsal centra keeled, anterior caudal centra with ventral pits, four metatarsals present. The hip has a deep pelvis with a long, slender ischium.

NEW DINOSAUR FOSSILS FROM LIGHTNING RIDGE, NEW SOUTH WALES

In recent years a number of new dinosaur bones have been recovered from the Lighting Ridge opal fields in northern New South Wales. These all come from the Griman Creek Formation (Early Cretaceous Age).

Many were collected by Dr Alex Ritchie and his team from an excavation known as the 'Sheepyard' site. The bones are undescribed at present, but include a diversity of types, mainly hypsilopho-dontid, ornithopod and theropods.

One large scapula (measuring 67cm) is from an ornithopod similar to *Muttaburrasaurus*.

Another large slab contains an associated collection of bones, including vertebrae, limb, rib and pelvic bones, probably from a medium sized theropod. These specimens are privately owned but are currently housed in the Australian Museum, Sydney.

▽ Reconstructed skeleton of *Muttaburrasaurus langdoni*, from Queensland

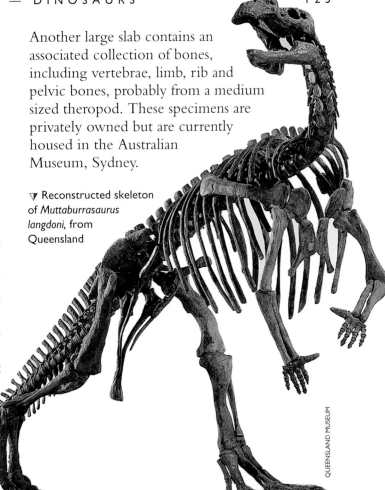

QUEENSLAND MUSEUM

NEXT PAGE reconstruction of *Muttaburrasaurus* by Peter Schouten

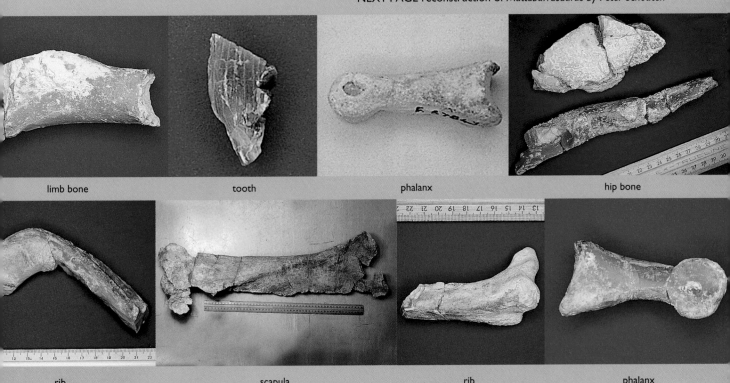

| limb bone | tooth | phalanx | hip bone |

| rib | scapula | rib | phalanx |

▲ A range of opalised dinosaur bones have been found from the Lightning Ridge opal fields, dated at about 100 million years old. The specimens shown here indicate the fauna included theropods, ornithopods and possibly thyreophorans. About a hundred or so of these bones are held in the collections of the Australian Museum in Sydney

FOOTPRINTS

The footprints of dinosaurs give us much additional information on the nature of Australian dinosaur faunas, and two sites in particular are very significant. One is near Winton, Queensland (Winton Formation, early Late Cretaceous), the other is from the coastline around Broome, Western Australia (Broome Sandstone, Early Cretaceous).

The Winton footprints are very well-preserved, and several weeks' work preparing them out by Dr Tony Thulborn, Dr Mary Wade, and numerous volunteers makes the site one of the best-preserved dinosaur track localities in the world. The footprints were made in a soft clay overlying a firm sandstone. Over 3300 footprints representing some 150 individual dinosaurs tell the story of what happened in 10 seconds of time about 100 million years ago. A herd of small, grazing plant-eaters (footprint taxon *Wintonopus*) and a large number of small coelurosaurs (footprint taxon *Skartopus*) were in close association when a large carnivorous dinosaur (footprint taxon *Tyrannosauropus*) cornered them against some rocky bluffs. The small dinosaurs had no choice but to run out past the meat-eater,

and their tracks indicate that they sprinted at top speed to get around their enemy. Speeds of up to 20 km are suggested by their strides (Thulborn and Wade 1979, 1984). The large carnivore had prints up to 75 cm in length, indicating a tyrannosaurid of large proportions, about 12 m or so in length. The footprints are now on public display at Lark Quarry on the road outside of Winton, Queensland.

Although dinosaur footprints have long been known from the Broome area, only over the past few years, through the diligence of Mr Paul Foulkes, a resident of Broome, working with Dr Tony Thulborn and Mr Tim Hamley, of the University of Queensland, has a large number of new dinosaur trackways of many different varieties been discovered throughout the area. These range as far to the north of Broome as outcrops of the Broome Sandstone are known. They indicate a diverse assemblage of at least ten different dinosaur species that inhabited the region about 110–120 million years ago. The footprints occur in two main environmental settings—one is a lagoonal setting, which is dominated by sauropod tracks; the other being a swampy forested setting, where theropods, ornithopods and thyreophorans dwelled (Thulborn *et al.* 1994)

▼ Large sauropod dinosaurs once inhabited the coastal plains near Broome, Western Australia, about 110 million years ago. By Michael Skrepnick

DINOSAUR FOOTPRINTS
From the Late–Early Cretaceous Winton Formation, Queensland

3–26 cm
Wintonopus latimorum

3–6 cm
Skartopus australis

up to 58 cm long
Tyrannosauropus sp.

VARIOUS DINOSAUR FOOTPRINTS
From Broome, Western Australia

21 cm long
?stegosaur hand

80 cm long
sauropod

17 cm long
Wintonopus

25 cm long
?stegosaur foot

46 cm
theropod

to 53 cm long
Megalosauropus broomensis

ORDER SAURISCHIA
SUBORDER THEROPODA

FOOTPRINT GENUS MEGALOSAUROPUS

SPECIES *Megalosauropus broomensis* Colbert and Merrilees 1967

AGE Early Cretaceous (Barremian–Aptian)

LOCALITY Gantheaume Point, near Broome, and other exposures of the Broome Sandstone to the north of Broome

Dinosaur footprints had long been known from Gantheaume Point, near Broome, in Western Australia. Although Ludwig Glauert had published a small note about them in 1952, it wasn't until 1967 that Dr Ned Colbert and Mr Duncan Merrilees undertook a field study and published a detailed report on them. They described footprints of a medium-sized carnivore similar to *Megalosaurus*, which they named *Megalosauropus broomensis*. The largest prints they measured were about 38 cm; since then new finds have shown that *Megalosauropus* prints up to 53 cm in length, with a stride of nearly 2 m, occur in the same sandstone. These trackways indicate that *Megalosauropus broomensis* was a meat-eating theropod that grew to approximately 9 m in length.

TECHNICAL DATA The footprints of *Megalosauropus broomensis* are characterised as bipedal trackways, with tridactyl prints up to 53 cm long, pace angulation about 140–160°, stride about 2 m. The prints have a digit angulation of about 35–45°, digit II being considerably longer than digits III and IV. Phalangeal formula as follows: 3 for digit II, 4 for digit III, 5 for digit IV.

MEGALOSAUROPUS BROOMENSIS

Sketch of the trackways exposed at the base of the cliff at Gantheaume Point, near Broome (as mapped by Colbert and Merrilees 1967). The different shades indicate separate trackways—the main trackway is shown in heavy stipple, The line of arrows shows direction of movement. The large arrow indicates north

1 metre

▼ Three-toed theropod print of *Megalosauropus broomensis*, approximately 53 cm in length

JOHN A. LONG

FOOTPRINT GENUS TYRANNOSAUROPUS

SPECIES Indeterminate

AGE Early Late Cretaceous (Cenomanian)

LOCALITY Lark Quarry, 120 km southwest of Winton, central Queensland (Winton Formation)

The trackway of *Tyrannosauropus* (meaning 'foot of *Tyrannosaurus*') is represented by 11 footprints, some of which show evidence of a sharp, pointed claw on the end of each toe. The footprints have a mean length of about 52 cm, suggesting an animal about 9–10 m long. Unlike those of *Megalosauropus*, the largest of which are of similar length, the *Tyrannosauropus* prints are broader and deeper, suggestive of a heavily-built, large animal, most likely something similar to *Tyrannosaurus*. The stride of the large predator was from 2.8 to 3.7 m, and the estimated speeds of the animal range from 5.3 to 8.5 km/h, suggesting it was using a walking gait. It took a slightly weaving course, with a tendency to slow down, as the first four strides are longer and deeper than the next four. The last two strides were very short and the animal then turned sharply right for some reason (Thulborn and Wade 1984).

TECHNICAL DATA From Thulborn and Wade (1984): large tridactyl footprints approximately 52 cm long, having three short digit impressions, which emerge from a deep basin-like 'pad' of the foot; each digit sharply defined, straight with a V-shaped tip and no indication of interphalangeal nodes. Digits II and IV distinctly shorter than digit III. The footprints often lack a clear outline of the back of the foot. The trackway is narrow with a mean pace angulation of about 170°.

FOOTPRINT GENUS SKARTOPUS

SPECIES *Skartopus australis* Thulborn and Wade 1984
Age Lower Late Cretaceous (Cenomanian)

LOCALITY Lark Quarry, 120 km southwest of Winton, central Queensland (Winton Formation)

Skartopus is named from two Greek words: 'skartes' ('nimble') and 'pous' ('foot'). The three-toed prints are almost bilaterally symmetrical and range in size from 2.9 to 5.7 cm, and in depth are sometimes as deep as 2 cm. The animal that made these tracks was probably a small coelurosaur—in the size range between 'bantams and half-grown emus', according to Thulborn and Wade (1984)—

▼ Large sauropod footprint, over 1 m wide, exposed in the shore platform at Broome, Western Australia

which was moving at a mean speed of 13 km/h. A measured sample of 34 of the track-makers concluded that the animals represented a range of normal size variation, and that the animals had a maximum hip height of about 22 cm. The fact that the *Skartopus* tracks are running in the opposite direction to the large tetanuran tracks suggests that there was a large herd of these animals trying to escape the predator (Thulborn and Wade 1979, 1984).

TECHNICAL DATA *Skartopus* is defined as follows: a small, digitigrade biped, with a pace angulation of 150°, and footprints that are tridactyl, slightly longer than broad, with distinct positive rotation. Digit imprints are narrow, straight and sharply pointed without phalangeal pads. Digit III is the longest; digits II and IV are about equal in length, and equally divergent from digit III (interdigital angles, 25° and 30°). The posterior margin of the footprint is an oblique line, running posterolateral to anteromedial, either straight or arched forwards. The ratio pace length to footprint length is between 5.2 and 9.1; the ratio of stride length to footprint length is between 10.6 and 17.3.

SUBORDER SAUROPODOMORPHA

FOOTPRINT GENUS INDETERMINATE

SPECIES Indeterminate

AGE Early Cretaceous (Neocomian–Barremian)

LOCALITY Beach exposures of the Broome Sandstone, around the Broome region

Large, rounded sauropod tracks, some up to 1.5 m long, are commonly found along the coastline at low tide, although these have often eroded away revealing the underneath layers or 'underprints' of squashed sedimentary rock below the original trackway. In some cases the prints are very deep, forming deeper stacks of disrupted sediment below them. Some of the best-preserved sauropod tracks include well-preserved hand (manus) prints immediately in front of the slightly larger foot (pes) prints (Thulborn *et al.* 1994).

TECHNICAL DATA At least two different types of sauropod tracks are recognised by Thulborn *et al.* (1994), the commonest form has footprints ranging between 45 and 120 cm, with rare occurrences over 1 m. The largest ones are

▼ Mr Paul Foulkes points to sauropod tracks exposed as weathered 'underprints', at Broome, Western Australia

JOHN A. LONG

bean-shaped and measure 1.5 m. In general this form has subcircular to elliptical footprints arranged in a wide gauge, zig-zag arrangement. Thulborn *et al.* (1994) believe that these trackways are referrable to the ichnogenus *Brontopodus*. Some trackways have hand impressions immediately in front and to the lateral margin of the footprint. One form has a subrectangular foot with a well-defined digit I, and represents a different ichnogenus. Until Thulborn's team completes their study we have no further information as to how many different types of sauropods were represented in the Broome Sandstone.

SUBORDER THYREOPHORA

FAMILY STEGOSAURIDAE

FOOTPRINT GENUS INDETERMINATE

SPECIES Indeterminate

AGE Early Cretaceous (Neocomian–Barremian)

LOCALITY Beach exposures of the Broome Sandstone, to the north of the Broome region

The combination of a stubby, five-fingered handprint in association with broad, three-toed footprints occurs at a site north of Broome. This hand and foot digit formula is suggestive of a primitive thyreophoran, possibly a *Stegosaurus*-like animal. It may constitute the only Australian record of the family Stegosauridae. The trackway

△ Unusual five-fingered dinosaur handprint associated with a three-toed footprint from near Broome, Western Australia—possibly from a stegosaur

JOHN A. LONG

occurs at a remote location, which was vandalised in 1996 when some of the prints were stolen. The identification of the trackway as a possible stegosaur (Long 1990, Dayton 1991) was based on information supplied to the author by Dr Tony Thulborn, based on his reconstruction of what a stegosaur trackway should look like. He based this on the hand and foot skeleton of *Stegosaurus*.

TECHNICAL DATA Only thyreophoran dinosaurs, and especially stegosaurids, have this combination of a five-fingered manus and a three-toed pes. No description of this trackway has been published, so further details are not available at this stage.

ORDER ORNITHISCHIA
SUBORDER ORNITHOPODA

FOOTPRINT GENUS WINTONOPUS

SPECIES *Wintonopus latimorum* Thulborn and Wade 1984

AGE Lower Late Cretaceous (Cenomanian)

LOCALITY Winton, central north Queensland (Winton Formation). Also *Wintonopus* sp. at Prices Point, north of Broome, Western Australia (Broome Sandstone)

Wintonopus latimorum is represented by over a thousand well-preserved prints at the Lark Quarry site. The genus name comes from the town of Winton, and the Greek word 'pous', meaning 'foot'; the species name is from the Latin word meaning 'stonemason', as a tribute to the many volunteers who assisted the Lark Quarry excavation. The prints are easily recognisable by their strong asymmetry, without a heel impression. Thulborn and Wade (1984) suggest that the animal that made these prints was a medium-sized ornithopod. Of 57 tracks analysed only one appeared to represent an animal larger than 1 m at the hip, in fact 81 per cent of them were less than 50 cm high at the hip. The animals moved at a fast, running gait at speeds of 12–20 km/h, without any evidence that they were slowing down.

In addition, a trackway at Prices Point, north of Broome, was discovered by Paul Foulkes, and later identified as *Wintonopus* sp. by Long (1990, 1993). This track has sometimes only two digits preserved, sometimes three, and has the same typical shape as the Winton tracks. However, at this stage it has not been formally described so is indeterminate as to its exact species.

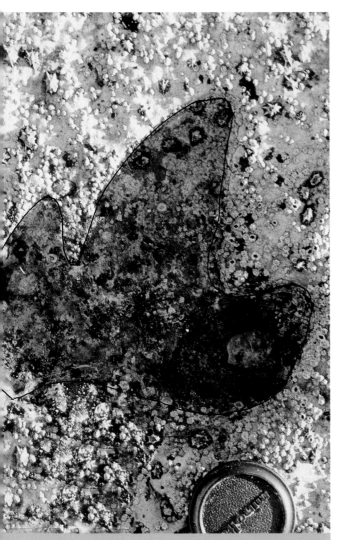

▲ *Wintonopus* footprint from north of Broome (about 17 cm wide)

TECHNICAL DATA *Wintonopus* is characterised by its narrow, digitigrade bipedal footprints, with a pace angulation of around 160°, and a footprint size of between 3 and 27 cm, but usually 7–8 cm long, and slightly wider than long, with a distinct positive rotation. Digit impressions are broad, with rounded to blunt tips, and lacking phalangeal pads. Digit III is longer than the others; digit IV is shorter and narrower than digit III; digit II, the shortest, is widely separated from digit III (interdigital angle, about 60°)—sometimes digits II and III are completely separated. The posterior margin of the foot is anteriorly convex. The ratio of pace length to foot length generally between 8 and 13.5 (or rarely within 4–15); ratio of stride length to footprint length is between 8 and 27, but usually within the range 16–24 (Thulborn and Wade 1984).

FOOTPRINT GENUS INDETERMINATE

SPECIES Indeterminate

AGE Early Cretaceous (Neocomian–Barremian)

LOCALITY Beach exposures of the Broome Sandstone, around the Broome region

There are several kinds of broad-footed tracks of probable ornithopod dinosaurs, some being large, clover-leaf-like feet, 45 cm or so in length, similar in size to a *Muttaburrasaurus*-type of animal. Current work on the tracks by Dr Tony Thulborn and his team will determine how many species are present.

JOHN LONG

▲ Large ornithopod dinosaur print, approximately 45 cm in length, from the Broome district, Western Australia

AUSTRALIAN AMPHIBIANS, MARINE REPTILES, CROCODILIANS, PTEROSAURS, BIRDS AND MAMMALS

INTRODUCTION

The other animals that lived alongside the dinosaurs are included here to complete the overall picture of the faunal communities in Australia during the Mesozoic Era. The last-known labyrinthodont amphibian on Earth, *Koolasuchus cleelandi,* was living in southern Victoria during the Early Cretaceous polar ecosystem (Warren *et al.,* 1997).

The seas were dominated by large swimming reptiles—the turtles, ichthyosaurs, mosasaurs and plesiosaurs—none of which are classified as dinosaurs seeing that all these groups stemmed off from the reptile family tree well before dinosaurs had evolved. Turtles, which belong to one of the most primitive of all reptile groups, the anapsids, are relatively unchanged today from their Mesozoic ancestors. Mosasaurs—poorly represented in Australia, however—were large sea-going lizards with flippers and powerful tails for propelling them under water. They are closely related to the living monitor lizards (or goannas, as they are called in Australia).

The ichthyosaurs and plesiosaurs are well-represented in Australia, by many isolated bones and some articulated, nearly complete skeletons. The ichthyosaurs were dolphin-like in appearance, and had long snouts armed with many small teeth for catching fishes. In some ways they were the dolphins of the reptile world and probably had similar lifestyles. The Plesiosauria has two major groups, the long-necked types with small heads (plesiosaurids) and the short-necked, robust types with long snouts (pliosaurids). Plesiosaurian remains occur in shallow marine and freshwater deposits, and are best known from the opal beds of South Australia and New South Wales, and from the marine Cretaceous of Queensland. Only two Mesozoic crocodiles have been recorded from Australia—one from the opal beds at Lightning Ridge (believed to be closely allied to the living saltwater crocodile), and another fairly complete but as yet undescribed form, from central Queensland.

The pterosaurs were flying reptiles that lived in the Mesozoic Era. The largest known pterosaur, *Quetzalcoatlus northropi,* had a wingspan of 12–15 m—larger than any bird that has ever lived. Well-preserved pterosaur fossils indicate that these creatures may have had a furry coat and consideration of their functional anatomy suggests that they were capable of active flight, rather than just gliding. Australia has a few insubstantial pterosaur remains.

Birds are represented in the Mesozoic of Australia by leg bones of a primitive flying bird from Queensland (*Nanantius*) and feathers, which could be from either birds or dinosaurs. These are preserved in fine mudstones from eastern Victoria.

Although mammals first appeared in the upper Middle Triassic, and were widely established by the Cretaceous Period, their presence in Australia at this time was only proven in the mid-1980s when a small jaw from Lightning Ridge was found. This find pushed back the record of mammals in Australia from about 30 million years ago (the previously oldest known mammal) to about 100–110 million years ago. Further recent finds in the 1990s have quadrupled the number of mammal jaws from the Cretaceous of Australia, but only two of the genera are represented by teeth well-enough preserved for the taxa to be formally named. Incidentally, mammals have recently been found in the intervening age range, from the latest Palaeocene (56 million years old) near Murgon, in Queensland, to fill in more of the gaps! Mammals evolved from synapsid reptiles at about the same time as the first dinosaurs appeared and, like dinosaurs and birds, must be considered as one of the 'specialised' subgroups of reptiles.

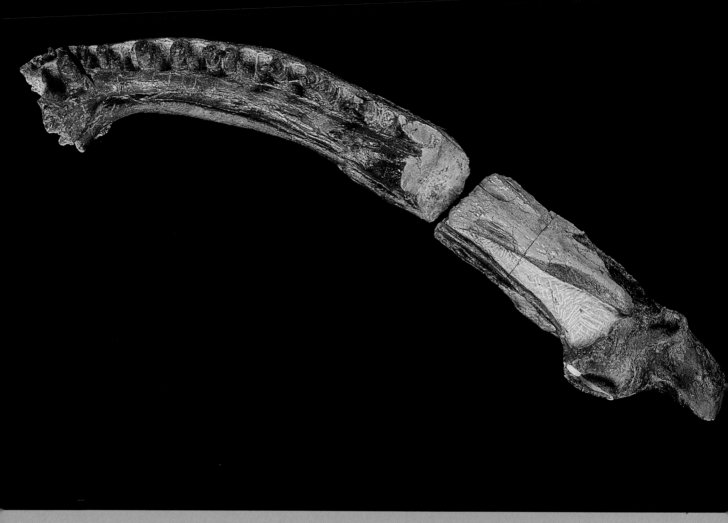

△ Lower jaw of a large labyrinthodont *Koolasuchus cleelandi* from the Early Cretaceous of Victoria (bar scale is 5 cm)

AN AMPHIBIAN

CLASS AMPHIBIA
SUBCLASS LABYRINTHODONTIA

FAMILY CHUGITOSAURIDAE

GENUS KOOLASUCHUS

SPECIES *Koolasuchus cleelandi* Warren *et al.* 1997

AGE Early Cretaceous (Aptian)

LOCALITY Coastal outcrops near Cape Paterson
through to Inverloch, East Gippsland, Victoria
(Strzelecki Group)

The first suggestion that labyrinthodonts had survived through to the Cretaceous Period came in the form of an unusual jawbone that could not be confidently attributed to a dinosaur or other reptile. The jaw was described by Jupp and Warren (1986), who concluded it belonged to a tem-nospondyl amphibian of uncertain affinities. Since then, other well-preserved lower jaws, plus shoulder girdle bones (clavicles), some skull fragments, and one partial skull of a large chugitosaurid labyrinthodont have been found along the East Gippsland coastal exposures. Some of these remains were described by Warren *et al.* (1991), giving indisputable proof that labyrinthodonts survived longer in Australia than anywhere else in the world, where most families died out at the end of the Triassic Period (about 100 million years earlier!). A complete description and formal naming of the new amphibian was recently published (Warren *et al.* 1997). It is named after Lesley Kool and Craig Cleeland. The giant amphibian from Victoria may have reached sizes of 3–4 m long, and probably occupied the niche usually taken by crocodiles, which are often common in Cretaceous river ecosystems, but were absent from the Victorian polar environment. The question of how these giant amphibians survived in such a cold climate is intriguing. However, just as giant Japanese salamanders survive winters in freezing cold streams, maybe our ancient labyrinthodonts developed similar adaptations.

TECHNICAL DATA A brachyopoid labyrinthodont having a lower jaw ramus in which the articular bone is excluded from the dorsal surface of the postglenoid area by a suture between the surangular and the prearticular. It is distinguished from *Siderops* and *Haddrokosaurus* by its absence of coronoid teeth.

MARINE REPTILES

SUBCLASS ANAPSIDA
ORDER CHELONIA

FAMILY DESMATOCHELYIDAE

This family unites the two genera of fossil turtles from Australia with the North American *Desmatochelys* and the European *Allopleuron*. Gaffney (1991) defines this family as having a plastron with very large central and peripheroplastral fontanelles, reducing the hypoplastral-hyoplastral contact to a narrow projection; the skull has the jugal missing a medial process, and the shoulder girdle has a wide scapular angle.

GENUS NOTOCHELONE

SPECIES *Notochelone costata* (Owen 1882)

AGE Early Cretaceous

LOCALITIES Throughout the outcropping Cretaceous marine rocks of Queensland in the crescent stretching from Hughenden to near Julia Creek, and south beyond Boulia (Toolebuc Formation and Allaru Mudstone). See Gaffney (1981) for a full listing of specimen sites.

Remains of this little turtle were first described by Sir Richard Owen in 1882 as a new genus 'Notochelys'; this name was later found to be already in use, however, and had to be changed to *Notochelone* (from the Greek words meaning 'southern turtle'). It is known from many fragments of shell (plastron), limb bones and three complete skulls; fragments of this turtle are also common throughout the marine Cretaceous rocks of Queensland. Gaffney (1981) accepted *Notochelone* as a valid genus of turtle, yet did not give a new diagnosis, on the basis of much unstudied new material. He did, however, illustrate the shell and gave a new restoration of it showing both the carapace (from dorsal view) and plastron (underneath, ventral view). *Notochelone* had a maximum length of about 1 m,

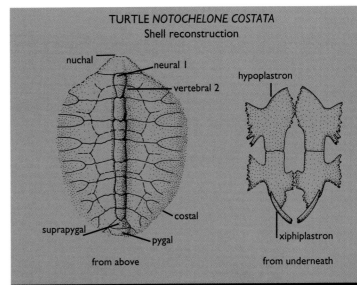

TURTLE *NOTOCHELONE COSTATA*
Shell reconstruction

△ Carapace of the extinct turtle *Notochelone costata*

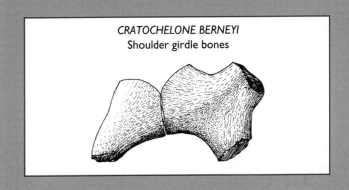

CRATOCHELONE BERNEYI
Shoulder girdle bones

and probably looked much like the marine green turtle of today, and probably had a similar lifestyle.

TECHNICAL DATA *Notochelone* has a carapace with a marked ridge along the back. The genus is definable by the distinct pattern of carapace and plastron bones.

GENUS CRATOCHELONE

SPECIES *Cratochelone berneyi* Longman 1915

AGE Early Cretaceous

LOCALITIES Sylvania Station, about 33 km from Hughenden, northern Queensland (?Toolebuc Formation)

Cratochelone is known from one specimen, comprising a shoulder girdle, incomplete front paddle bones (parts of the humerus, ulna and radius) and pieces of the shell (plastron), indicating it was a very large sea-going turtle with a maximum size of about 2.25 m, comparable with the giant fossil turtles from North America (for example, *Archelon*). The name *Cratochelone* comes from the Greek meaning 'giant turtle', and the species name honours Mr F. Berney, who gave the bones to the Queensland Museum (Longman 1915). In his review of Australian fossil turtles Gaffney

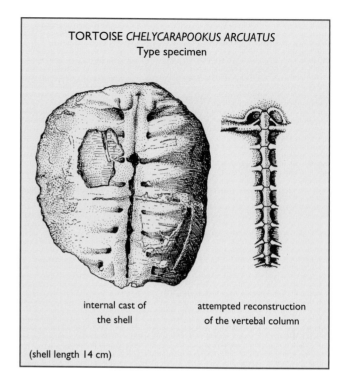

TORTOISE *CHELYCARAPOOKUS ARCUATUS*
Type specimen

internal cast of the shell

attempted reconstruction of the vertebal column

(shell length 14 cm)

(1981) noted that the material of *Cratochelone* cannot be related to other known fossil or living turtles, and we must await new discoveries before the affinities of this unusual, large turtle can be determined.

TECHNICAL DATA *Cratochelone* is defined by its robust, short scapulo-precoracoid and coracoid

⚠ The fossil turtles inhabiting the inland seas of Australia during the Cretaceous Period were not unlike the living marine turtles such as this hawksbill, *Eretmochelys imbricata*

natural size

ulna

phalanx

▲ Mosasaur paddle bones from Gingin, Western Australia

JOHN A. LONG

bones, and its large size. According to Longman (1915, p.25), the plastral plate cannot be associated with any of the known fossil turtles treated in the literature, so on this basis he erected the new genus. Gaffney (1981) adds that, after he had examined the specimen, it 'permits no useful comparison'. However, he suggests it is closer to the protostegids than to other chelonian groups, and reinforces this view in his 1991 review paper.

GENUS CHELYCARAPOOKUS

SPECIES *Chelycarapookus arcuatus* J.W. Warren 1969

AGE Early Cretaceous

LOCALITIES Near Casterton, western Victoria (Merino Group)

This little tortoise lived in fresh water, and is known from the internal cast of the shell collected by a Mr J. Macpherson. The genus name comes from the Greek 'chelys' 'turtle', and the district of Carapook in western Victoria; the species name refers to the arcuate depressions at the front of the shell on its inside face. The specimen was collected early this century and first briefly described by Frederick Chapman (1919). Dr Jim Warren, formerly of Monash University, looked at the specimen in detail and named it as a new form, *Chelycarapookus*, in 1969. In a review of Australian fossil turtles, Dr Gene Gaffney (1981, 1991) suggested that *Chelycarapookus* could be a cryptodire, or long-necked tortoise, because it lacked the pleurodiran condition of having a fused pelvis.

TECHNICAL DATA Unusual features of the genus are the presence of an excavation, or fossa, just anterior to the first rib on each side, and that the neural elements of the carapace become wider posteriorly. The necks of the posterior ribs are also very wide, and fuse further from the vertebral column than in any other chelonian (Molnar 1991).

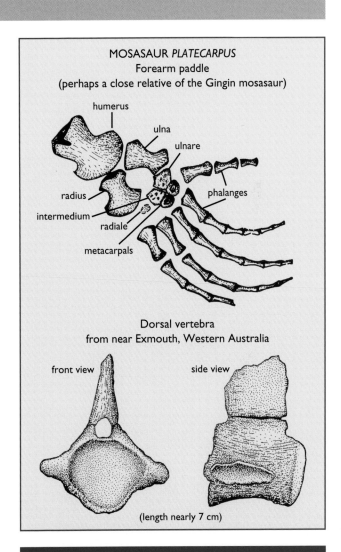

MOSASAUR *PLATECARPUS* Forearm paddle (perhaps a close relative of the Gingin mosasaur)

humerus, ulna, ulnare, radius, intermedium, radiale, metacarpals, phalanges

Dorsal vertebra from near Exmouth, Western Australia

front view   side view

(length nearly 7 cm)

SUBCLASS DIAPSIDA
ORDER SQUAMATA

FAMILY MOSASAURIDAE

Mosasaurs were large, predatory marine lizards related to the modern-day goannas (varanids). Their limbs evolved into paddles, but their long tails propelled them in the water; in this they were unlike plesiosaurs, which relied on their powerful flippers for propulsion. The only two records of mosasaurs in Australia are both from the Late Cretaceous of Western Australia.

GENUS INDETERMINATE

SPECIES Indeterminate

AGE Late Cretaceous (Turonian–?Santonian)

LOCALITY MacIntyre's Gully, near Gingin, Western
Australia (Molecap Greensand)

Three small bones were collected in 1956 by a student, Mr S. St Warne, who was studying the geology of the Gingin area. The bones consist of a left ulna and parts of other paddle bones, enabling Lundelius and Warne (1960) to suggest that the material closely resembles *Platecarpus* from the Upper Cretaceous of Kansas. Unfortunately, as the ulna is partially abraded the angles for articulation of the other fin bones cannot be accurately determined. The other paddle bone is one of the finger bones (phalanges), and this agrees with *Platecarpus* and *Clidastes* in its slender proportions, rather than with the more advanced mosasaurs, which had many squat bones in the paddles.

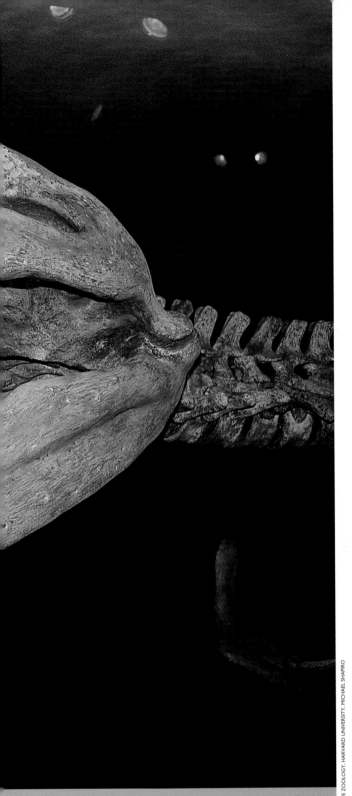

△ Close up detail of the skull of *Kronosaurus queenslandicus*, as reconstructed in the Museum of Comparative Zoology, Harvard University, USA

MUSEUM OF COMPARATIVE ZOOLOGY, HARVARD UNIVERSITY, MICHAEL SHAPIRO

TECHNICAL DATA The Gingin mosasaur ulna is 5.4 cm in length, compared with 8.8 cm for a large *Platecarpus* ulna (total length of 4.3 m for *Platecarpus*, Russell 1967). From this comparison, I estimate an approximate length for the Gingin mosasaur of 2.6 m.

GENUS INDETERMINATE

SPECIES Indeterminate

AGE Late Cretaceous (Late Maastrichtian)

LOCALITY Giralia Range, south of Exmouth Gulf, Western Australia (Miria Formation)

Another record of mosasaurs in Western Australia comes from the end of the Cretaceous. Three dorsal vertebrae were found in close association, as part of the same individual, by Mr Peter Bindon, from the Western Australian Museum, and myself in July 1991. They are well-preserved and show the neural arches complete on two specimens. They represent an animal of estimated total length of about 6–8 m, but are not identified at this stage to any specific mosasaur family.

TECHNICAL DATA The Giralia mosasaur vertebrae have strongly convex anterior central faces and posteriorly they are strongly concave. They are dorsals, one well-preserved specimen showing large dorsolateral rib facets.

ORDER PLESIOSAURIA

FAMILY PLIOSAURIDAE

Pliosaurs were short-necked, robust plesiosaurs, often with long snouts. They were wholly marine reptiles whose powerful flippers propelled them along in the water. I follow Brown (1981) in including the families Rhomaelosauridae and Polycotylidae within the Pliosauridae.

GENUS KRONOSAURUS

SPECIES *Kronosaurus queenslandicus* Longman 1924

AGE Early Cretaceous (Aptian–Albian)

LOCALITY The type specimen came from near Hughenden, central western Queensland (Toolebuc Formation). Other material, including the near-complete skeleton owned by Harvard University, came from the Army Downs region and Grampian Valley, about 50 km north of Richmond (Wallumbilla Formation).

The photograph of the reconstructed skeleton of *Kronosaurus* in the Harvard University's Museum of Comparative Zoology immediately brings to mind the classic picture of a giant sea monster of the dinosaur age. A full 12.8 m long, the skeleton represents possibly the largest marine reptile that ever lived. Its skull is 2 m long. The teeth reach up to 30 cm in length, the crowns of which were about 12 cm long, the remainder forming the root. The name *Kronosaurus* comes from the Greek god Kronos, who ate his children. The interesting story of the discovery of this remarkable specimen is related in chapter 3. It took some 25 years before preparation was completed. Mr Ted White began this work, and spent the first three years working on the skull. Little else was done until the 1950s when Al Romer managed to get a sponsor, Mr Godfrey Cabbot, who had an interest in sea serpents, to pay for the mount-ing of the whole skeleton. This job was completed by Jim Jensen and Arnold Lewis, who incorporated much original bone material into the finished mount, which went on public display in 1959 (Romer and Lewis 1960). In reconstructing this skeleton, many missing parts, like the front paddles, and missing parts of the skull, were restored after similar pliosaurids. Heber Longman, who described and named *Kronosaurus* from fragmentary material in 1924 and in 1930, was never to see the

COLIN MCHENRY

▲ Excavating new material of *Kronosaurus* in the field, near Hughenden, Queensland

▼ Close-up of reconstructed front paddle of *Kronosaurus queenslandicus*

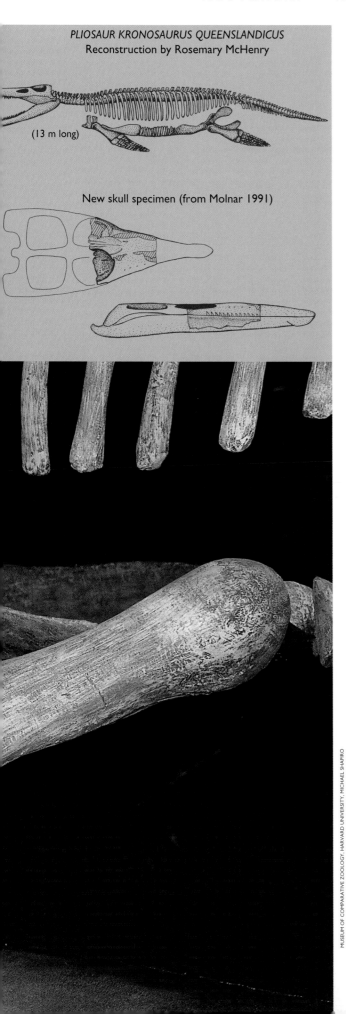

PLIOSAUR KRONOSAURUS QUEENSLANDICUS
Reconstruction by Rosemary McHenry

(13 m long)

New skull specimen (from Molnar 1991)

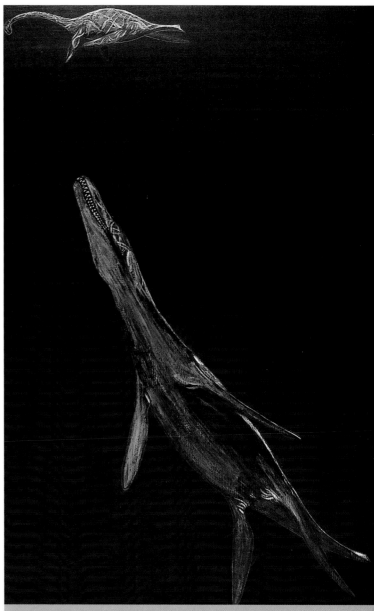

▲ Reconstruction of the 13-metre long pliosaur *Kronosaurus queenslandicus* by Rosemary McHenry

reconstructed specimen on display in the USA. The Harvard specimen, even though it confirmed Longman's earlier interpretations about *Kronosaurus* being a pliosaur, is still to be studied in detail so we do not actually have a description of the skeleton, despite its relative completeness.

Molnar (1991) doubts that the Harvard skeleton is really the same species as the type material described as *Kronosaurus queenslandicus* by Longman (1924), since the two specimens come from different aged strata (Molnar 1982a, 1982b, 1991). New material of *Kronosaurus* is currently being studied at the University of Queensland by Mr Colin McHenry, who has exmained the

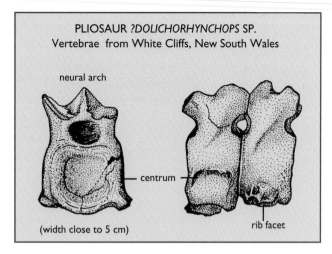

PLIOSAUR *?DOLICHORHYNCHOPS* SP.
Vertebrae from White Cliffs, New South Wales

neural arch

centrum

(width close to 5 cm)

rib facet

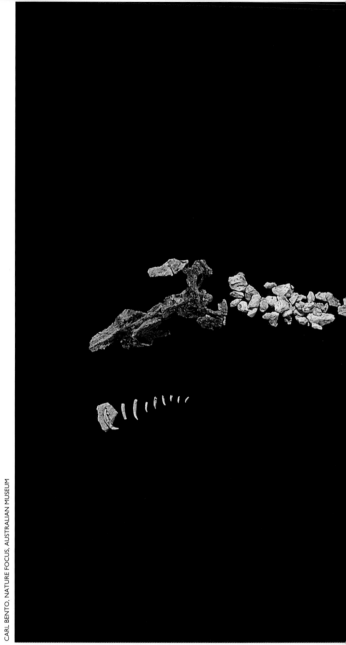

Harvard specimen along with much new material from Queensland. McHenry has seen no evidence so far that more than one taxon is present (McHenry pers. comm. 1997). *Kronosaurus*, like other pliosaurs, had powerful jaws for preying on turtles and elasmosaurs, as remains of these have been found in its gut contents (McHenry, pers. comm. 1997). In recent years a number of new finds of *Kronosaurus* remains have been made by teams from the Queensland Museum and the University of Queensland.

TECHNICAL DATA The original type material of *Kronosaurus*, along with new skulls from the same strata (Toolebuc Formation), indicates that the skull was streamlined, with large adductor fossae for the jaw muscles. The orbits faced forwards, upwards, and sideways. There is a prominent mid-nasal ridge, which, behind the orbits, continues to rise in a moderately high parasagittal crest (formed by the parietals), which runs along the long axis of the skull. There is no evidence for the high transverse hump on top of the squamosals that was included in the 1958 Harvard mount of the skeleton (McHenry, pers. comm. 1997).

GENUS ?DOLICHORHYNCHOPS

SPECIES Indeterminate

AGE Early Cretaceous (Albian)

LOCALITY White Cliffs, northern New South Wales (Coreena Formation)

The remains of this reptile were first described by Etheridge in 1897 as a new species of *Cimoliasaurus*, *C. leucospelus*, from opalised bones collected at White Cliffs. Persson (1960) reviewed Australian plesiosaur material and decided that the remains were too incomplete to be assigned to a new species but that they were similar

enough to the American *Dolichorhynchops* in tooth and vertebrae shape to be tentatively assigned to that genus. The Australian material of *?Dolichorhynchops* consists of teeth, backbones and fragments of flipper bones. Other vertebrae of pliosaurid plesiosaurs have also been recorded from Weatherby, near Richmond, Queensland, by Persson (1960) but are too incomplete for positive identification.

TECHNICAL DATA Dr A. Cruickshank informed me that '*Dolichorhynchops*' is synonymous with *Trinacromerum*, to which it should be assigned. As determination of *Dolichorhynchops* or *Cimoliasaurus* really depends on skull morphology, it is difficult to assign these specimens to any genus.

△ Opalised skeleton of 'Eric', a small *Leptocleidus* species from Cooper Pedy, South Australia

▽ The opal fields of Cooper Pedy, South Australia, have yeilded some spectacular fossil finds

The cervical vertebrae figured here actually resemble those of *Leptocleidus* sp. (A. Cruick-shank, pers. comm. 1996).

GENUS LEPTOCLEIDUS

SPECIES *Leptocleidus clemai* Cruickshank and Long 1997a

AGE Early Cretaceous (Hauterivian–Barremian)

LOCALITY North of Kalbarri, Western Australia (Birdrong Sandstone)

SPECIES *Leptocleidus* sp.

AGE Early Cretaceous (Aptian–Albian)

LOCALITY Coober Pedy, South Australia (Maree Formation)

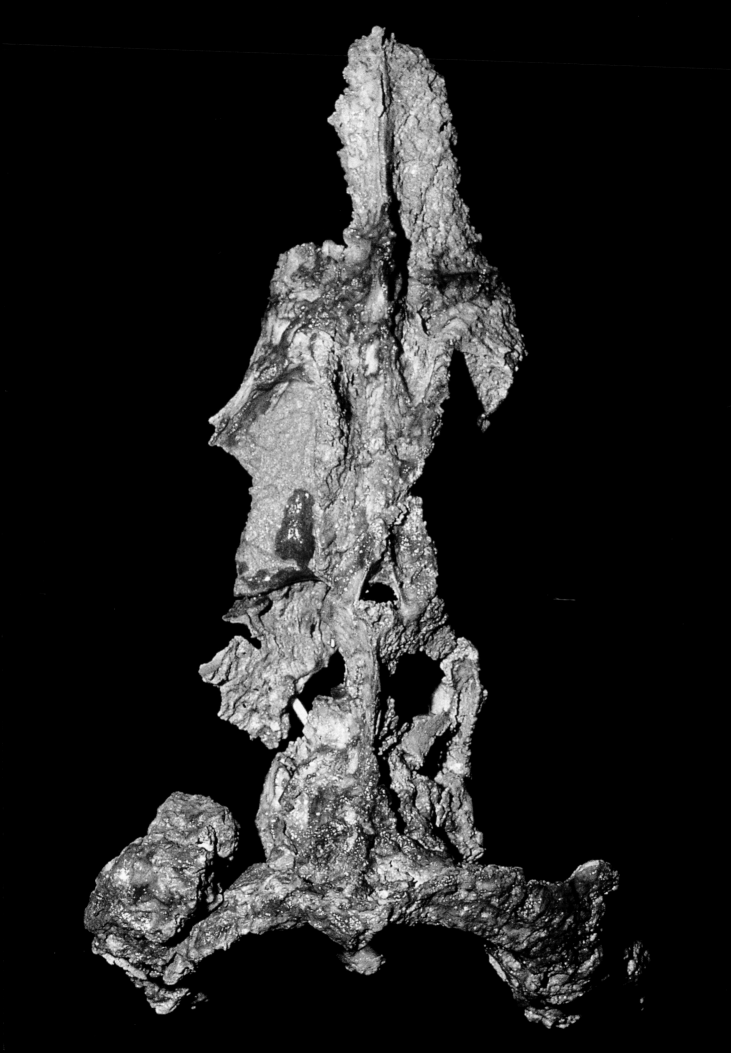

▽ ▲ Femur (below) and vertebrae (above) of the small pliosaurid *Leptocleidus clemai* from near Kalbarri, Western Australia

◁ Skull of the *Leptocleidus* sp. 'Eric' from Coober Pedy, South Australia

The genus *Leptocleidus* was first described from a relatively complete skeleton found in the Wealden beds of England, of Hauterivian– Barremian age, and has recently been identified from near Port Elizabeth, South Africa, by Dr A. Cruickshank (*L. capensis*). The Western Australian specimens of *Leptocleidus clemai* come from north of Kalbarri, where they had weathered out of the top metre of the Birdrong Sandstone. In 1991 geology students Glyn Ellis, Ian Copp and Greg Milnar located a string of reptile vertebrae weathering out of the sandy bank. After several field trips, three partial skeletons were recovered, consisting of several hundred bones—mostly vertebrae but also limb bones (femur, humerus, ulna, radius, tibia), and parts of the shoulder girdle and pelvic girdle, and rib fragments. *L. clemai* is named in honour of the person who partly sponsored the field-work, Mr John Clema. It is the largest known *Leptocleidus*, being probably close to 3 m long—about 15–20 per cent larger than the type species and significantly larger than 'Eric', the species from Coober Pedy. The Kalbarri *Leptocleidus* specimens all represented partially decomposed carcasses, which probably floated for a while, losing the head and the ends of the paddles to scavengers. The remains then sank in shallow marine conditions, as indicated by the *Teredo*-bored petrified wood and fragmentary ammonites which occur in the same layer.

The famous 'Eric' specimen was acquired by the Australian Museum in 1993 and has an interesting background, well told by Ritchie (1990). It was found by an opal miner and came out in thousands of small fragments, which were then meticulously prepared and reconstructed by Paul Willis. The specimen is an almost complete skeleton replaced by opal, comprising the skull, fragment of lower jaw, several teeth, most of the vertebral column, ribs, parts of the shoulder girdle and pelvic girdle, and most of the pectoral and pelvic limb bones, but it is missing the smaller paddle bones. 'Eric' is now known to belong in the genus *Leptocleidus* and probably represents a new species (N. Schroeder, 1997 thesis, Monash University). It was a small pliosaurid slightly less than 2 m long, and fed on fish: remains of teleost vertebrae have been found inside its gut along with rounded pebbles (gastroliths) that were either used as an aid to digestion or for ballast. *Leptocleidus* probably lived near shore and occupied a similar niche to the one occupied by seals today.

TECHNICAL DATA The following list of diagnostic features for the genus is taken from Cruickshank and Long (1997b). *Leptocleidus* is a small genus of pliosaurid; it has a skull that is triangular in outline, with a prominent midnasal ridge which merges with the parasagittal crest, flanked by deep grooves or depressions, which in turn cause the orbital rims to stand up from the general profile of the skull. The tooth count is reduced to 21 positions on each side of the upper jaw (5 in each premaxilla, 16 in each maxilla).

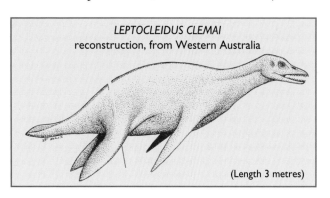

LEPTOCLEIDUS CLEMAI
reconstruction, from Western Australia

(Length 3 metres)

No complete lower jaw is known for the genus, but an estimated tooth count is 35 (based on *L. capensis*). The pectoral girdle is primitive, having large clavicles and interclavicles and small scapulae; the humerus has a very much more symmetrical (fan-shaped) distal end. None of the vertebrae are compressed, the cervicals being spool-shaped and the neural arches relatively large; the cervical vertebral count is in excess of 13. *Leptocleidus* differs from early forms in one other feature: it possesses a forward-pointing expansion ('cockscomb') on the squamosal mid-line, at the rear of the parasagittal crest, on the back of the skull.

▽ New pliosaurid from Queensland
▷ Opalised plesiosaur paddle from Andamooka, South Australia

NEVILLE PLEDGE

QUEENSLAND MUSEUM

GENUS INDETERMINATE

SPECIES Indeterminate

AGE Early Cretaceous

LOCALITIES Lightning Ridge, northern New South Wales (Griman Creek Formation)

A large propodial element was collected from Lightning Ridge some years ago. The bone is of good quality opal and was used by a conman as collateral to secure large financial loans. After the criminal was captured by police, the bone eventually ended up as a registered specimen in the Australian Museum (F 102462), Sydney. The bone is from a pliosaurid similar to *Leptocleidus clemai* from Western Australia, but significantly larger. As it is missing its proximal end, its exact identification may never be known.

SUPERFAMILY PLESIOSAUROIDEA

FAMILY ELASMOSAURIDAE

Elasmosaurs are the well-known, slender, long-necked forms of plesiosaurs, which have about 72 cervical vertebrae in the neck. They probably did not raise their heads out of water but projected them forwards, held straight underwater whilst swimming (Dr A. Cruickshank, pers. comm. 1996).

GENUS WOOLUNGASAURUS

SPECIES *Woolungasaurus glendowerensis* Persson 1960

AGE Early Cretaceous (Albian)

LOCALITY From near 'Rainscourt', Richmond district, Queensland (Wallumbilla Formation)

SPECIES ?*Woolungasaurus* sp.

AGE Cretaceous (stratigraphic layer uncertain)

LOCALITY Neale's River, northwest of Lake Eyre, South Australia (Maree Formation)

Woolungasaurus is named from an Aboriginal word 'woolunga' meaning a mythical reptile animal, and 'sauros' (Greek for 'lizard'). *Woolungasaurus* is one of Australia's best-represented plesiosaurs, being known from an incomplete skeleton consisting of some 46 vertebrae (mostly centra only), ribs, shoulder girdles, forearm (flipper) bones and partial hind flipper bones from Queensland, and a series of 12 vertebrae from inland South Australia. The South Australian specimen has the vertebrae with the same proportions as the Queensland form and, although incomplete, is therefore provisionally identified as the same genus (Persson 1960). The neck was very long and slender. A complete plesiosaur paddle which was excavated at Andamooka and later sold to a museum in St Paul, Minnesota, USA, has been referred to *Woolungasaurus* sp. by Pledge (1980).

A skull referrable to *Woolungasaurus* was described by Persson in 1982. A remarkable story surrounds this discovery. Part of the skull was first found in the bed of Yamborra Creek, near Maxwelton, by a Mr Noonan, donated to the

Queensland Geological Survey and later acquired by the Queensland Museum. In July 1976 the site was revisited by Dr Ralph Molnar, Dr Tony Thulborn and Dr Mary Wade; they found the front half of the same skull, along with some other vertebrae. As Ralph Molnar was then working in Sydney, that specimen became registered in the collections of the Australian Museum. Dr Alex Ritchie recognised that the front half of the skull they had found belonged with the back half which had been found earlier and sent to Dr Persson in Sweden. He then contacted Persson with the news. The skull is badly crushed, but because the associated neck vertebrae are well-preserved the specimen can be identified as *Woolungasaurus*. The skull is typical for an elasmosaur, having a slender

QUEENSLAND MUSEUM

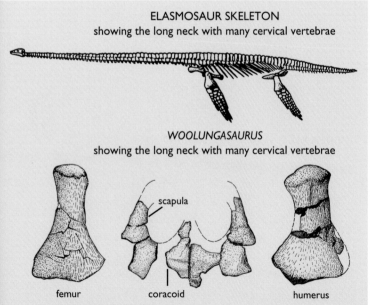

ELASMOSAUR SKELETON
showing the long neck with many cervical vertebrae

WOOLUNGASAURUS
showing the long neck with many cervical vertebrae

scapula

femur   coracoid   humerus

snout with about 40 sharp, slightly recurved teeth. If complete it would have measured about 42 cm long, giving the overall size for the beast at about 9.5 m, based on the relative skull to body length of *Hydrothecrosaurus*. In 1997 *Woolungasaurus* adorned an Australian stamp.

TECHNICAL DATA The diagnostic features of *Woolungasaurus* are seen in the cervical vertebrae which at the front of the neck are longer than they are broad, and at the rear of the neck are broader than long. The humerus is also longer than the femur. *Woolungasaurus* is believed by Persson (1982) to be closely related to *Hydralmosaurus*, from North America, as both these forms have an unusual thickening on the midline of one of the shoulder bones (coracoid).

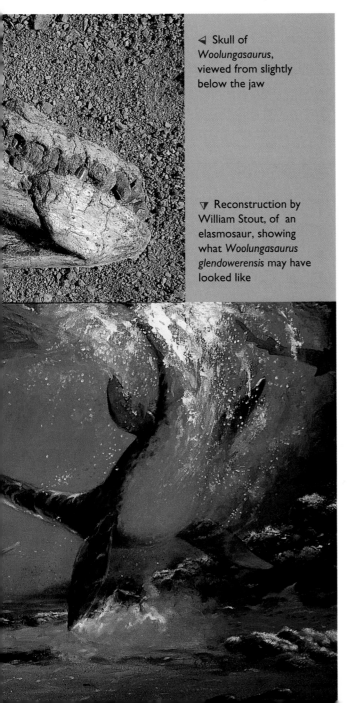

◁ Skull of *Woolungasaurus*, viewed from slightly below the jaw

▽ Reconstruction by William Stout, of an elasmosaur, showing what *Woolungasaurus glendowerensis* may have looked like

GENUS INDETERMINATE

SPECIES Indeterminate

AGE Late Cretaceous (Cenomanian–?Santonian)

LOCALITY Cooks deposit, just west of the township of Dandaragan, Western Australia (Molecap Greensand)

A number of isolated bones from plesiosaurs have been found from the Dandaragan and Gingin areas of Western Australia (Teichert and Matheson 1944, Long and Cruickshank 1998). One large plesiosaur dorsal vertebra from Dandaragan (Western Australian Museum no. 86.5.1) measures 11 cm in centrum diameter, and is probably an elasmosaurid, based on its size and elongated proportions. The specimens from Gingin and Dandaragan represent Australia's youngest known plesiosaurian fossils.

GENUS AND SPECIES INDETERMINATE

SPECIES Indeterminate

AGE Early Cretaceous (Albian)

LOCALITY Casuarina Beach, 2 km northwest of Darwin, Northern Territory (Darwin member, Bathurst Island Formation)

Although ichthyosaurs had long been known from the Casuarina beach site near Darwin, remains of plesiosaurs are much rarer from the site, and were first recognised in the 1980s by Peter Murray from the Northern Territory Museum, who collected and described them (Murray 1985, 1987). The beach outcrop makes finding fossil bones difficult because of encrustations by bryozoans, worms, oysters and algae. The recognisable plesiosaur bones include fragments of large limb bones (femur, humerus), vertebrae and a part of the pelvis (ilium). The elongate nature of the neck vertebra suggests that it comes from an elasmosaur, but its general proportions differ from the Queensland form *Woolungasaurus* in that the neck vertebra is more elongated and narrower. Murray (1987) estimates the Darwin elasmosaur to have been in the order of 8 m long.

FAMILY CIMOLIASAURIDAE

Cimoliasaurs were small, long-necked plesiosaurs, but are more primitive than elasmosaurs in having a shorter neck, a more robust build, different neck vertebrae proportions, and single-headed ribs (Persson 1960).

PLESIOSAUR
CIMOLIASAURUS
MACCOYI
Bones from White Cliffs,
New South Wales

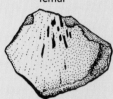

femur

part of humerus

body vertebra in
different views

GENUS CIMOLIASAURUS

SPECIES *Cimoliasaurus maccoyi*
Etheridge 1904

AGE Early Cretaceous (Albian)

LOCALITIES White Cliffs, northern New
South Wales (Coreena Formation)

Cimoliasaurus maccoyi was a small plesiosaur about 3–4 m long, whose bones are preserved partly as opal. First described as *Cimoliasaurus* by Etheridge in 1897, it was identified as a new species by Etheridge (1904) following the discovery of further material. It is based on some well-preserved neck vertebrae and some paddle bones. Other remains of this species include one tooth, fragments of ribs and flipper bones. The genus *Cimoliasaurus* is not unique to Australia; more complete material is known from England, France, North America and New Zealand. TECHNICAL DATA Several workers for example, David Brown and Sam Welles, do not consider the genus *Cimoliasaurus* to be valid as it was erected on scrappy material (McHenry pers. comm 1997). The Australian species, however, was diagnosed by Persson (1960) based on the absence of longitudinal ridges on the neck vertebrae, and the general proportions of the vertebrae, which have flat centra. The femur is also quite slender. Persson suggested that the Australian species may even be the same as *Cimoliasaurus planus* from England and France.

ORDER ICHTHYOSAURIA

FAMILY STENOPTERYGIIDAE

The stenopterygiid ichthyosaurs were characterised by having a modified flipper bone arrangement with broad attachment of the fins to the body.

GENUS PLATYPTERYGIUS

SPECIES *Platypterygius longmani* Wade 1990

AGE Early Cretaceous (Albian)

LOCALITIES Headwater of the Flinders River, near O'Connell Creek and near Julia Creek, Telemon

NEVILLE PLEDGE

▲ Skull and front of body skeleton of the ichthyosaur *Platypterygius longmani* from Queensland

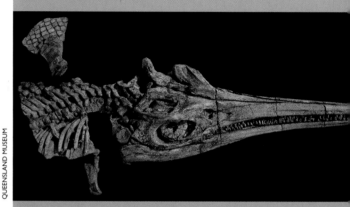

QUEENSLAND MUSEUM

▼ Reconstruction by Brian Choo of the seven-metre long ichthyosaur *Platypterygius longmani*

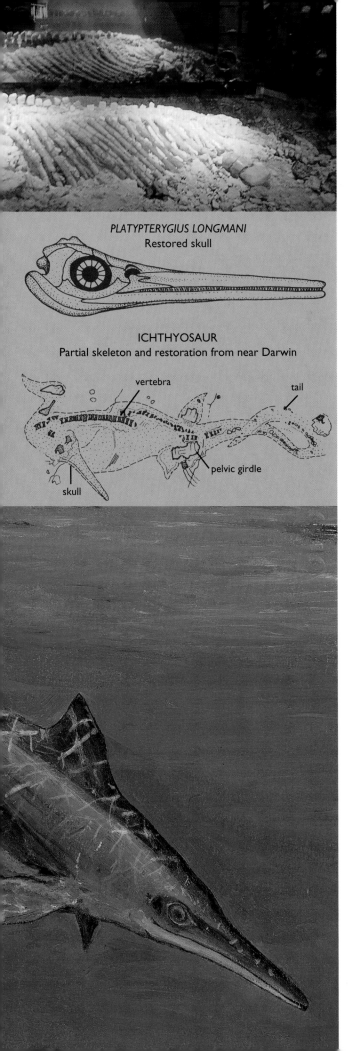

PLATYPTERYGIUS LONGMANI
Restored skull

ICHTHYOSAUR
Partial skeleton and restoration from near Darwin

vertebra

tail

skull

pelvic girdle

Station; Galah Creek, and other localities in north central Queensland (Toolebuc Formation and Allaru Mudstone)

The history of early finds of *Platypterygius* (then known as *'Ichthyosaurus australis'*) has been outlined in chapter 3. Since the early discoveries, which were studied by McCoy, Etheridge and Longman, several new specimens of ichthyosaurs have been found in Queensland; but it was when Dr Chris McGowan reviewed Cretaceous ichthyosaurs in 1972 that he decided *'Ichthyosaurus australis'* was actually referrable to *Platypterygius* (meaning 'flat wing', alluding to the flipper). Dr Mary Wade (1984, 1990) gives an historical summary of the events leading to the sorting out of the different material attributed to *Platypterygius*. Wade (1990) erected the new species *Platypterygius longmani* to include all the Queensland ichthyosaurs previously labelled as *Platypterygius (Ichthyosaurus) australis*.

The most complete specimen came from Telemon Station, collected by Edgar Young and donated to the Queensland Museum in 1934 (there is a photograph of this excavation in Wade 1990). It is a skull, part of the front flippers and most of the backbone, measuring 5.6 m in length. Larger isolated bones indicate that *Platypterygius* reached a maximum length of about 7 m. Wade (1984) suggests that *Platypterygius* was rather dolphin-like in its lifestyle, and could probably sit on the bottom of the sea-floor to rest for short periods. Like all ichthyosaurs, *Platypterygius* was propelled by its powerful tail. The shape of the animal is very conservative, along regular ichthyosaur lines: long snout, large eyes, and broad front flippers. The structure of the front flippers of *Platypterygius* is quite interesting, as there are many rows of tightly fitting digits. Thus, the fingers have many more bones than normal, but the wrist has been simplified with bones actually lost.

TECHNICAL DATA The species is defined by Wade (1990) in a lengthy diagnosis, so only some of the more distinctive features are listed here. *Platypterygius longmani* was a moderately large, long-snouted ichthyosaur with many teeth on the premaxilla and maxilla; with a prenarial maxillary foramen present, and with oval to bean-shaped nares, and an oval-shaped orbit; the supraoccipital is large, enclosing most of foramen magnum. The atlas-axis complex is heart-shaped in end view. The neural arches are strongly inclined from axis to 32nd vertebrae; from 11th to 20th neural arches, the crests of the neural spine are divided into anterior and posterior peaks by a V-shaped apical notch. The 46th vertebra is the first with a single rib articulation; small tailfin vertebrae have dorsolateral ridges on either side of the neural groove. The femur is 70 per cent as long as the humerus, which has prominent dorsal and ventral trochanters. Three primary fingers lie one below each of the radiale, intermedium and ulnare, supported by three anterior (radiale) accessory fingers, and three posterior (ulnare) fingers. These fingers form a close-fitting pavement of rectangular phalanges. All fin blade bones are tightly appressed.

GENUS PLATYPTERYGIUS

SPECIES Indeterminate

AGE Early Cretaceous (upper Albian)

LOCALITIES Fanny Bay, Nightcliff and Casuarina Beach, near Darwin (Darwin, Northern Territory Member, Bathurst Island Formation)

Ichthyosaur bones were first discovered in the Darwin region in 1915 by workmen, and this was reported anonymously in 1924 (in the Australian Museum magazine). In recent years a number of new finds have been made; although many are of isolated vertebrae, two specimens at Nightcliff are partial skeletons of ichthyosaurs. Some of the skull and fin skeletons are preserved, but not enough to determine the genus. Murray (1985) suggested that the Darwin ichthyosaurs are close to *Platypterygius* in general proportions and vertebral structure, and in his subsequent paper (Murray 1987) he was able to confirm this identification with additional material of the humerus and scapula. The Nightcliff ichthyosaur is interpreted as being a rotten carcass which floated in the sea, being scavenged on by other creatures, until the headless, limbless, bloated carcass became stranded on the beach and was eventually buried by sediment. Reconstructed, the body of the Nightcliff ichthyosaur would have been slightly longer than 2 m, and based on the isolated bones from the various sites around Darwin, the maximum length of these beasts would have been up to 4 m.

GENUS PLATYPTERYGIUS

SPECIES Indeterminate

AGE Early Cretaceous (Valanginian–Hauterivian)

LOCALITIES Giralia Anticline, Cardabia Station, east of Coral Bay, and near Kalbarri, both in Western Australia (Birdrong Sandstone)

In 1993 an expedition from the University of Western Australia had picked up some bones of both plesiosaurians and ichthyosaurs from newly discovered outcrops of the Birdrong Sandstone (McLaughlin *et al.* 1995). This led a Western Australian Museum party there the following year, when a partially articulated skeleton of an ichthyosaur was found by myself, as well as a fragmented skull of a second specimen by Ms Kristine Brimmell. The remains of the first skeleton include about 20 poorly preserved vertebral centra, fragments of the humeri and ribs. From his initial study of these remains Brian Choo, a University of Western Australia honours student, has identified the skeleton as belonging to *Platypterygius*. Choo believes it is closer to the European species than to the Queensland *P. longmani*. I have collected isolated ichthyosaur vertebrae—from the Birdrong Sandstone in the Kalbarri district—which are very large, in fact comparable in size with those of adult *P. longmani*.

GENUS ?PLATYPTERYGIUS

SPECIES Indeterminate

AGE Late Cretaceous (Cenomanian–?Santonian)

LOCALITY Cooks deposit, just west of the township of Dandaragan, Western Australia (Molecap Greensand)

Isolated large vertebrae from ichthyosaurs were found and illustrated from Dandaragan, Western Australia by Teichert and Matheson (1944). Their large size and compact shape makes them very similar to those of *Platypterygius*, to which they are tentatively referred here, as further suggested by Wade (1990, p.136). These are all of Late Cretaceous age and constitute the last-known ichthyosaurs from Australia, but are still slightly older than the last known forms from Europe (Coniacian).

Isolated ichthyosaur vertebrae that were reported in earlier editions of this book as coming from the Miria Formation, Giralia Range, have since been identified as shark vertebrae; thus, no ichthyosaurs are recorded from this formation.

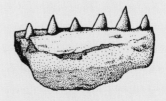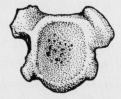

CROCODILE *?CROCODYLUS SELASLOPHENSIS*
Jaw fragment and vertebra
from Lightning Ridge, New South Wales

▲ The living saltwater crocodile, *Crocodylus porosus*, grows to more than 7 metres long. Crocodiles have a fossil record spanning back some 230 million years

CROCODILIANS

SUBCLASS ARCHOSAURIA
ORDER CROCODYLOMORPHA

FAMILY ?CROCODYLIDAE

This family includes the living crocodiles. The fossil remains of crocodiles found at Lightning Ridge are too incomplete to be confidently assigned to this family. A new, well-preserved specimen from Queensland is Australia's most complete fossil crocodile, but only preliminary comments about it were available at the time this book was being compiled. In addition to these crocodilians, Molnar (1991) also reports that part of a crocodile skull or lower jaw has been found in the Toolebuc Formation near Hughenden, which is currently unstudied.

GENUS ?CROCODYLUS

SPECIES *?Crocodylus selaslophensis* Etheridge 1917

AGE Lower Cretaceous

LOCALITY Lightning Ridge, New South Wales (Griman Creek Formation)

Etheridge (1917) described a fragment of an opalised crocodile jaw as a new species, *Crocodylus selaslophensis*, even though there are not enough features seen in the specimen to warrant erecting a new species. Over the years several new specimens of crocodile bones, including neck vertebrae, have turned up from the same locality, and these were described by Molnar (1980c). Other remains include a sacral and a tail vertebra, a neck rib, and partial right thigh bone (femur) and two incomplete left shin bones (tibiae). Molnar suggests that the material is not referrable to *Crocodylus*, but definitely represents a eusuchian, the group containing modern crocodilians. The presence of a crocodilian with procoelous vertebrae in the Early Cretaceous of Australia means that, along with *Hylaeochampsa* from England, and specimens from western Europe attributed to *Crocodylus* which require restudy—together they represent the oldest occurrence of modern crocodiles (Molnar 1980c).

TECHNICAL DATA Molnar (1980c) does not believe that the type material of *C. selaslophensis* is *Crocodylus* but leaves the specimens under Etheridge's *C. selaslophensis* pending further, more complete remains. The neck vertebrae are procoelous (the front face is deeply concave) as in modern crocodiles. The teeth of the dentary are set in sockets in a groove, a feature found in few other Mesozoic forms, and in one Eocene form, *Eocaiman*, from Argentina.

GENUS INDETERMINATE

SPECIES Indeterminate

AGE Late Cretaceous (Cenomanian)

LOCALITY Near Isisford, central Queensland (Winton Formation)

A partial skeleton of a neosuchian crocodilian has recently been found in a large sandstone nodule from central Queensland. It is Australia's most complete Mesozoic crocodile found to date, comprising the sacral and posterior dorsal vertebrae, femora, ribs, dorsal armour and parts of the pelvis and the fibula, and represents an animal about 1.5 m in length.

TECHNICAL DATA The armour on its back is interesting because it has four longitudinal rows of scutes posteriorly and six rows anteriorly, and the scutes suture with their lateral neighbours. They lack anterior processes and do not imbricate. Molnar and Willis (1996) suggest that the crocodile is related to *Bernissartia*, from Belgium, but is more advanced in having six rows of scutes in the mid region.

PTEROSAURS

The pterosaurs were the winged flying reptiles that lived at the same time as the dinosaurs. Pterosaur remains from Australia are very rare, only a handful of specimens being known. Besides the Queensland and Western Australian specimens listed below there are some bones from the Early Cretaceous terrestrial sandstones of southern Victoria, but so far these have not been studied in detail (Rich and Rich, 1989).

ORDER PTEROSAURIA
SUBORDER PTERODACTYLOIDEA

FAMILY ?PTERANODONTIDAE

A pteranodontid pterosaur is characterised by its generally toothless beak, its large head with well-developed crest, and large size. Although the Queensland pterosaur jaws have teeth, Molnar and Thulborn (1980) use the broader definition of the family, and include the new specimen in this family because of close similarities in the shoulder girdle to *Ornithocheirus*.

GENUS ?ORNITHOCHEIRUS

SPECIES Indeterminate

AGE Early Cretaceous

LOCALITY Near the Hamilton River, western Queensland (Toolebuc Formation)

The first remains of a pterosaur, or flying reptile, from Australia were found in marine limestones in Queensland only in the late 1970s, being represented by a shoulder girdle, one vertebra, and a section of the fused lower jaws. Unlike many pterosaur remains in other countries, that are crushed flat, the Queensland material was prepared out of limestone using acetic acid, and so preservation is three-dimensional. Molnar and Thulborn (1980) believe the new specimen compares closely with the English pterosaur *Ornithocheirus*, but as both this form and the Queensland material are incomplete a confident identification

⮝ Reconstruction by Emily Dortch of the pterosaur *Anhanguera*, a genus that may also have inhabited Queensland and New Zealand

cannot be made without further finds. The Queensland pterosaur had a long beak armed with teeth for catching fish and had an estimated wingspan of around 2–4 m.

TECHNICAL DATA This pterosaur has a beak with teeth, indicating it was closer to *Ornithocheirus*. The vertebra is almost hollow with fine struts supporting it. The scapulocoracoid is V-shaped with scapular and coracoid wings enclosing an angle of about 75°; in general form it approaches the shape of the scapulocoracoid in Late Cretaceous forms like *Nyctosaurus* or *Pteranodon*.

GENUS ?PTERANODON

SPECIES Indeterminate

AGE Early Cretaceous

LOCALITY Near the Hamilton River, 70 km east of Boulia, western Queensland (Toolebuc Formation)

This is a further specimen from the Toolebuc Formation, consisting of an incomplete pterosaur pelvis, which was described by Molnar (1987) as being very close to that of *Pteranodon*. The specimen was very well-preserved, showing the hip socket in natural position, indicating the legs did not function in the same way that birds' legs do and suggesting that the pterosaur was not capable of bipedal walking. *Pteranodon* was a large form, known from several species, some of which had wingspans of up to 9 m. The only doubt expressed by Molnar (1987) as to whether the specimen is definitely *Pteranodon* or not is the associated presence of the other pterosaur remains described above, which all appear to be from the same-sized animal. If they did indeed all come from one species, the beast in question would, by its teeth, clearly differ from *Pteranodon*.

TECHNICAL DATA The pelvis which resembles that of *Pteranodon* includes much of the right ilium, most of the right pubis, some of the prepubis and a small part of the right ischium, with remnants of two fused sacral centra and a sacral rib. The pelvis exactly matches in size and shape that of *Pteranodon ingens* (Molnar 1987).

FAMILY ANHANGUERIDAE

This family of pterosaurs includes some well-preserved species of the genus *Anhanguera* from South America. These have a prominent crest on the snout and lower jaws, and have separate scapular and coracoid bones (that is, not fused).

GENUS cf. ANHANGUERA

SPECIES Indeterminate

AGE Cretaceous

LOCALITY near Hughenden, north-central Queensland (Allaru Mudstone)

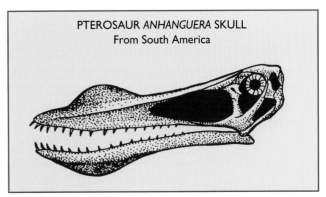

PTEROSAUR *ANHANGUERA* SKULL
From South America

The specimen consists of the snout of a pterosaur missing the tip and broken through the hole in front of the eyes where the nostrils were located. It shows teeth present. *Anhanguera* is a well-preserved pterosaur known from the Santana Formation of Brazil. It had a 4 m wingspan and, in addition to its crested snout, it bore a similar crest underneath the lower jaws, believed to have aided in stabilising the beak when fishing in full flight.

TECHNICAL DATA The snout of *Anhanguera* is distinct in having a low, rounded crest. However, a full description by Molnar and Thulborn, outlining the diagnostic features of the specimen, is yet to be published (to appear in the British journal *Palaeontology*).

FAMILY ?AZHDARCHIDAE

The azhdarchids were the largest of all pterosaurs, characterised by their long necks and special neck vertebrae structure, but also by some features of the limb skeleton.

GENUS INDETERMINATE

SPECIES Indeterminate

AGE Late Cretaceous (Late Maastrichtian)

LOCALITY Toothawarra Creek, Giralia Range, south of Exmouth Gulf, Western Australia (Miria Formation)

AUSTRALIAN CRETACEOUS PTEROSAURS
from the Toolebuc formation pterosaur
from western Queensland

shoulder girdle

restored pelvis

In 1990 I noticed an unusual bone in the fossil collections of the Western Australian Museum. It had been found over thirty years earlier by Mr Eric Carr, a museum worker, in rocks of Late Cretaceous age (about 66–70 million years old). The bone had a slender shaft, leading me to think that it could not be one of the marine reptiles, like a turtle or plesiosaur, as these have short, stocky limb bones. Casts of the bone were sent out for second opinions and the verdict soon returned that it was the proximal end of an arm bone (an ulna) of a pterosaur. The Western Australian specimen appears to be closely related to the North American giant *Quetzalcoatlus*. This was a gigantic pterosaur from Texas, which has a 12–15 m wingspan, making it the largest creature to have ever flown. The Western Australian pterosaur had an estimated wingspan of 3.8 m (based on *Quetzalcoatlus*) or 4.9 m (based on *Pteranodon*), making it the largest Australian pterosaur, and one of the last pterosaurs in the world.

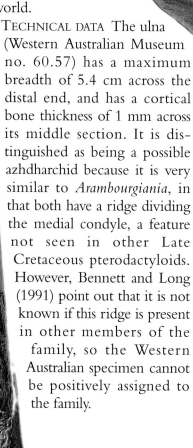

KRIS BRIMMELL

TECHNICAL DATA The ulna (Western Australian Museum no. 60.57) has a maximum breadth of 5.4 cm across the distal end, and has a cortical bone thickness of 1 mm across its middle section. It is distinguished as being a possible azhdharchid because it is very similar to *Arambourgiania*, in that both have a ridge dividing the medial condyle, a feature not seen in other Late Cretaceous pterodactyloids. However, Bennett and Long (1991) point out that it is not known if this ridge is present in other members of the family, so the Western Australian specimen cannot be positively assigned to the family.

◁ An arm bone (partial ulna) from a pterosaur from Western Australia which had an estimated wingspan of up to 5 m natural size

BIRDS

The presence of birds in the Cretaceous of Australia is based on a few small bones from Queensland. Fossil feathers from Victoria may be interpreted as being from either birds or, possibly, dinosaurs.

FAMILY INDETERMINATE

Enantiornithines were small flying birds which achieved widespread distribution in the Cretaceous, also being known from South America, Madagascar, Mexico, and Mongolia. They were true flying birds, unlike any older birds or bird-like dinosaurs (including *Archaeopteryx),* which may have had poor flying ability. As they are based on scant remains, families have not been sufficiently defined at this stage.

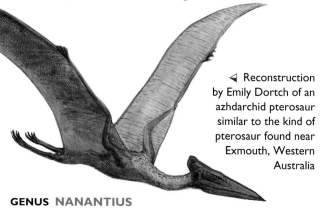

◁ Reconstruction by Emily Dortch of an azhdarchid pterosaur similar to the kind of pterosaur found near Exmouth, Western Australia

GENUS NANANTIUS

SPECIES *Nanantius eos* Molnar 1986

AGE Lower Cretaceous (Albian)

LOCALITY Northeast paddock of Warra Station, near Hamilton Hotel, and Canary Station, both west Queensland (Toolebuc Formation)

Nanantius is the only known Mesozoic bird named from Australia, and was found by Dr Ralph Molnar by dissolving a sample of bone-rich Toolebuc Formation limestone in acid, then picking through the bone fragments in the residues. *Nanantius* comes from the Greek words 'nanos', meaning 'dwarf', and 'en-antios', meaning 'opposite', alluding to the class Enantiornithes. The species name means 'dawn' and refers to the fact that this is the oldest known occurrence of this subclass of birds. It is represented by leg bones, about 3 cm long (the tibiotarsus), and a cervical vertebra which shows that *Nanantius* was the size of a European blackbird (Molnar 1986, Kurochkin and Molnar 1997). Aside from the Jurassic *Archaeopteryx*, *Nanantius* could possibly be one of the oldest true flying birds. Molnar (1986) suggests that *Nanantius* shows links between the Australian and South American Cretaceous vertebrate faunas. TECHNICAL DATA *Nanantius* is defined by the following characters: its very narrow, slender tibiotarsus; the distinct ascending process of the astragalus (seen by the groove on the tibiotarsus); the distal end of the tibiotarsus has a very narrow intercondylar sulcus, with the medial condyle more nearly cyclindrical than for any other described enantiorninthine. There is also a marked depression anterior to the fibular crest proximally.

GENUS INDETERMINATE

SPECIES Indeterminate

AGE Early Cretaceous (Aptian/Albian)

LOCALITY Road cutting near Koonwarra, near
  Leongatha, eastern Victoria (Strzelecki Group)

The only other evidence of birds in the Mesozoic of Australia are feathers preserved in fine Early Cretaceous mudstone from Koonwarra road cutting, in east Gippsland, Victoria. To date, about five such

feathers have been found, and these are from a small bird about the same size as *Nanantius*. The feathers were first described by Talent *et al.* (1966), and have also been reported by Waldman (1970) and Rich and van Tets (1982). Whether these feathers belong to birds or dinosaurs is now an open question, since the 1996 discovery of a small, feathered dinosaur from the Early Cretaceous of China.

MAMMALS

Over the last few years the record of fossil mammals known from Australia has greatly increased, adding more monotreme taxa as well as the first possible placental mammal. These rare but important discoveries indicate that during the Cretaceous Period the major groups of mammals had spread out right across the Earth, and because placental mammals are not present in the mid-late Tertiary, it is assumed that they either died out (possibly with the dinosaurs) or re-entered Australia at a later stage (Rich *et al.* 1997).

CLASS MAMMALIA
SUBCLASS MONOTREMATA

FAMILY STEROPODONTIDAE

This family is defined on one specialisation within monotremes, according to the recent analysis of Flannery *et al.* (1995)—having a deep dentary. It includes only one taxon, *Steropodon galmani*.

GENUS STEROPODON

TYPE SPECIES *Steropodon galmani* Archer *et al.* 1985

AGE Lower Cretaceous (Albian)

LOCALITY Lightning Ridge, New South Wales
  (Griman Creek Formation)

The discovery of this little jaw caused a scientific sensation as it was the first record of a Mesozoic mammal known from Australia, and followed three decades of intense searching by American and Australian scientists. The specimen was purchased through opal dealers David and Alan Galman by the Australian Museum. The name comes from two Greek words—'sterope', meaning 'flash of lightning', and 'odous', meaning 'tooth', alluding to the opalised preservation; the species name honours the Galman brothers. Although only just under 3 cm in length, the jaw is beautifully preserved as a pseudomorph in opal, and shows three well-preserved molar teeth. The

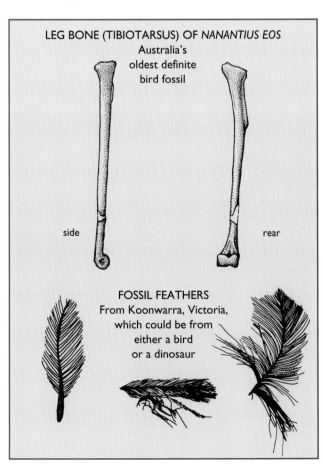

LEG BONE (TIBIOTARSUS) OF *NANANTIUS EOS*
Australia's oldest definite bird fossil

side   rear

FOSSIL FEATHERS
From Koonwarra, Victoria, which could be from either a bird or a dinosaur

nature of the complex teeth shape indicates a close relationship with the platypus, based on similarities shared with the milk teeth in baby platypus (Archer *et al.* 1985). It is also of interest in that it is one of the largest known Mesozoic mammals (along with *Kollikodon*), being about the size of a cat.

TECHNICAL DATA The teeth of *Steropodon* are technically diagnosed as differing from other monotreme teeth by the retention of a cristid obliqua and talonid basin, and the possession of a pre-entocristid. It differs from all other mammals that are not monotremes in having just three lower molars, and in the very compressed trigonid anteroposteriorly, transverse metacristid, and a well-developed, wide talonid (Archer *et al.* 1985).

FAMILY KOLLIKODONTIDAE

This family was erected for the single specimen of *Kollikodon* and is based on the defining features of that species (listed below). The family is believed to be more primitive than the Steropodontidae because of the presence of four molar teeth, the loss of the fourth molar uniting *Steropodon* with all other monotremes.

GENUS KOLLIKODON

TYPE SPECIES *Kollikodon ritchiei* Flannery *et al.* 1995

AGE Lower Cretaceous (Albian)

LOCALITY Lightning Ridge, New South Wales (Griman Creek Formation)

This was the second record of a Mesozoic mammal known from Australia, and was discovered one decade after *Steropodon*. The name comes from the Greek words 'kollikos', meaning 'bun-like', and 'odous', meaning 'tooth', alluding to its bun-like molars (in fact, an early suggestion from Mike Archer for a name for this beast was 'hotcrossbunodon'!). The species name honours Dr Alex Ritchie who procured the specimen for the Australian Museum from opal miners in Lightning Ridge, along with two other Cretaceous mammal jaws lacking teeth.

It is about 3.6 cm in length, and, like *Steropodon*, consists of a lower jaw with three well-preserved molars and the holes (alveoli) for the premolar. The molars are broad and robust, lacking vertical cutting blades, with four rounded lophs, giving them the superficial appearance of hot-cross buns. *Kollikodon* was platypus-sized, slightly larger than *Steropodon*, and may have used

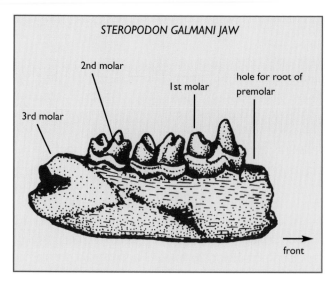

its usual teeth for crushing prey such as clams or snails. The large mandibular canal in the jaw suggests that *Kollikodon*, like other monotremes, had a bill, or at least a set of electrosensory organs at the front of its face. The large size of both *Steropodon* and *Kollikodon* may be a reflection of the cold weather southeastern Australia experienced at that time.

TECHNICAL DATA *Kollikodon* is unique amongst monotremes in having at least four molars; these have bunodont morphology and lack vertical cutting blades or crests. The lower jaw first molar protoconid is smaller than the metaconid; first molar (M1) is much smaller than the second molar (M2), and the posterior molars are more square-shaped than rectangular.

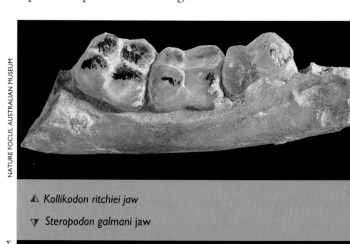

△ *Kollikodon ritchiei* jaw
▽ *Steropodon galmani* jaw

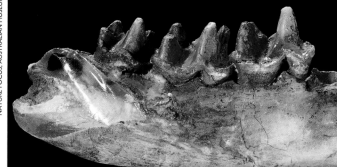

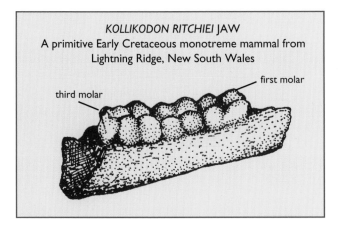

KOLLIKODON RITCHIEI JAW
A primitive Early Cretaceous monotreme mammal from
Lightning Ridge, New South Wales

third molar

first molar

▲ Jaw of Ausktribosphenos

GENUS INDETERMINATE (cited Rich et al.1997)

SPECIES Indeterminate

AGE Early Cretaceous (Albian)

LOCALITY Dinosaur Cove, western Victoria (Otway Group)

The discovery of a humerus belonging to a monotreme from the Dinosaur Cove site was reported in Rich *et al.* (1997, figure 1) as a second record of an early Cretaceous mammal from Victoria. As yet the specimen (MSC 011, Monash Science Centre) is undescribed.

**SUBCLASS THERIA
INFRACLASS PLACENTALIA
ORDER AUSKTRIBOSPHENIDA**

FAMILY AUSKTRIBOSPHENIDAE

GENUS AUSKTRIBOSPHENOS

SPECIES *Ausktribosphenos nyktos* Rich et al. 1997

AGE Early Cretaceous (Aptian)

LOCALITY Coastal exposures at Flat Rocks, Bunarong Marine Park, near Iverloch, eastern Victoria (Wonthaggi Formation, Strzelecki Group)

The discovery of this little jaw, only 16 mm long, by Nicola Barton in March 1997 turned theories on mammalian biogeography upside down. The jaw, containing four teeth, is identified as an early placental mammal, implying that both placentals and marsupials may have been widespread on all landmasses of the Earth by the Cretaceous. Alternatively, because of its highly unusual tooth morphology, Rich *et al.* (1997) suggest that *Ausktribosphenos* may represent an entirely new lineage of Mesozoic mammal that forms a sister group to the placentals.The name means 'southern three-coned (tooth)'. The word 'tribosphenic' describes a general pattern of early mammal teeth in which the molars have a triangular-shaped occlusal surface

with the apex towards the tongue. The species name 'nykos' is Greek for 'night', alluding to the long time the animal spent in darkness each year due to the extreme southern latitude of the fauna. *Ausktribosphenos* probably ate insects and small vertebrates living in the cool polar forests. Many species of insects and other arthropods have been described from the same sedimentary succession that produced the jaw of *Ausktribosphenos* and the diverse Victorian dinosaur fauna. The jaw (MSC 007) resides at the Monash Science Centre in Melbourne.

Dr Michael Archer (1997) holds a differing opinion to Tom Rich and his colleagues on the affinities of this little jaw. He believes that *Ausktribosphenos* may be a primitive sister taxon to the rest of the monotremes, and is thus more closely related to *Steropodon* and *Kollikodon*.

TECHNICAL DATA *Ausktribosphenos* is characterised as a placental by its postcanine dental formula, and it is distinguished from all other mammals in having a special, unnamed third cristid on the lower molar. It differs from monotremes in having a paraconid on the lower first molar, expanded molar trigonids and a tribosphenic wear pattern on the lower molars; it differs from all marsupials in having the lower posteriormost premolar with all three trigonid cusps developed, and three (not four) lower molars, and at least four lower premolars; it differs from all placentals in having a remnant of the surangular facet, a hypoconulid located close to the entoconid on the lower molars; and a crest on the lower first and second molars linking the hypoconulid and metaconid buccal to the entoconid, which is separately linked to the metaconid. It differs from a number of other early fossil mammals by details listed in Rich *et al.* (1997).

Ausktribosphenos was a small shrew-like mammal

eight

NEW ZEALAND DINOSAURS, MARINE REPTILES AND PTEROSAURS

Dinosaurs were first recognised from New Zealand in 1981; since then only a handful of isolated bones have been found. The fossil animal fauna of this age is found from three main regions. In the North Island, boulders containing fossil bones erode from the Tahora Formation and are found in the Mangahouanga Stream on the east coast. These have yielded a treasure trove containing the rare, scant remains of New Zealand's only dinosaurs, plus well-preserved remains of marine reptiles, including good skulls and partial skeletons of mosasaurs and plesiosaurs. In the South Island, the region from Waipara River in North Canterbury, stretching north to Haumuri Bluff, has yielded many superb specimens of mosasaurs and plesiosaurs, and has long been known as a rich area for vertebrate fossils. Similar finds also occur in the exposed beach rocks at Shag Point, near Otago. All of these sites are shallow marine deposits.

The question of how the dinosaurs survived the cold climate of Late Cretaceous New Zealand has been discussed at length by Molnar and Wiffen (1994), who reach no solid conclusions. They dismissed the 'hot-blooded dinosaur' theory on 'insufficient grounds' and argue against the gigantothermy hypothesis—according to which the large body size of some dinosaurs could have insulated them from the cold climates. They point out that this strategy relies, in part, on annual seasonal migration of animals to warmer climes, which would have been impossible in an island habitat.

The small size of the New Zealand dinosaurs and the cold climate appear to be good evidence, taken in conjunction with the Victorian polar dinosaur fauna, that some small dinosaurs must have had some internal heat-generating mechanism, even if it wasn't exactly the same, or as efficient as the hot-blooded metabolism of modern mammals and birds. Given the close relationship now established between birds and dinosaurs, it seems most probable that some dinosaurs had this capability.

MARINE REPTILES

SUBCLASS	ANAPSIDA
ORDER	CHELONIA

FAMILY PROTOSTEGIDAE

GENUS INDETERMINATE

SPECIES Indeterminate

AGE Late Cretaceous (Campanian–Maastrichtian)

LOCALITY Mangahouanga Stream site, North Island (Maungataniwha Sandstone)

Wiffen (1981) described bits of a fossilised turtle shell (plastron and carapace fragments), this being the only record of the group from the Mesozoic of New Zealand. It was suggested the eight turtle fragments were from a small member of the family Protostegidae, this being the first Southern Hemisphere record of the group.

SUBCLASS	DIAPSIDA
ORDER	SQUAMATA

FAMILY MOSASAURIDAE

GENUS MOSASAURUS

SPECIES *Mosasaurus mokoroa* Welles and Gregg 1971

AGE Late Cretaceous (Maastrichtian)

LOCALITY Haumuri Bluff, Cheviot and Waipara district, South Island (Laidmore Formation)

SPECIES *Mosasaurus flemingi* Wiffen 1990a

AGE Late Cretaceous (Campanian–Maastrichtian)

LOCALITY Mangahouanga Stream site, North Island (Maungataniwha Sandstone)

The well-known European genus *Mosasaurus* is represented in New Zealand by the species *M. mokoroa* from the Haumuri Bluff site on the South Island and *M. flemingi* from the North Island. Of these, *M. mokoroa* is the better represented of the two.

▷ The Mangahouanga Stream site, where New Zealand's scant dinosaur and pterosaur remains have all been found

DR EWAN FORDYCE

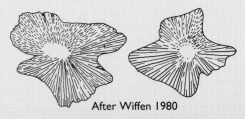

PROTOSTEGID TURTLE SHELL FRAGMENTS
From the Late Cretaceous of New Zealand

After Wiffen 1980

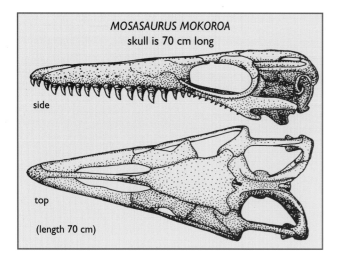

MOSASAURUS MOKOROA
skull is 70 cm long

side

top

(length 70 cm)

The holotype is a well-preserved disarticulated skull and one vertebra, and a second specimen consists of part of the left forelimb paddle. The skull measures some 70 cm in length, and was carefully removed from its limey sandstone casing by acetic acid preparation. *M. mokoroa* would have been about 10–12 m in length. *M. flemingi* is known only from the holotype specimen, comprising parts of the braincase, the jaw articulation (quadrate), four neck vertebrae and a tooth. It was a medium-sized species of *Mosasaurus*.

TECHNICAL DATA *M. mokoroa* is characterised by having a skull with a moderately concave prefrontal bone, and nearly straight frontal border; its quadrate has a suprastapedial process extending below the middle of that bone, and constricted medially to less than half its distal width; the infrastapedial process projects posterodorso-medially, and the stapedial pit is tilted posteriorly, with a deep vertical groove on internal face below the pit (Welles and Gregg 1971).

M. flemingi is characterised by its broad, robust, rectangular quadrate bone, which is similar to that of the typical *Mosasaurus* and yet has an angularity and prominence of the ventro-interior corner that is more like the condition seen in *Clidastes*. The suprastapedial process is large, and there are no foramina that pass through the tympanic cavity and main shaft (as in *Mosasaurus, Tylosaurus, Platecarpus* and *Clidastes*).

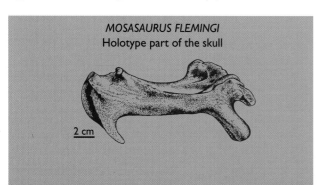

MOSASAURUS FLEMINGI
Holotype part of the skull

2 cm

GENUS MOANASAURUS

SPECIES *Moanasaurus mangahouangae* Wiffen 1980

AGE Late Cretaceous (Campanian–Maastrichtian)

LOCALITY Mangahouanga Stream site, North Island (Maungataniwha Sandstone)

The holotype specimen, which comprises a disarticulated skull, some vertebrae, paddle and rib bones, was found by Mr M. Wiffen on 12 October 1974. It was lying in a very large boulder, showing only three water-worn bones exposed on the surface. It had to be blasted in half to be removed. Unfortunately, one half of the specimen fell into a deep waterhole in the stream and could not be recovered until March 1978. Eventually, when the specimens were prepared from the rock using acetic acid, the parts of the skull could be reconstructed. The name *Moanasaurus* comes from the Maori word 'moana', meaning 'sea', and the Greek word 'sauros', meaning 'lizard', and the species name reflects the locality of the find.

Wiffen (1990a) assigned some additional material belonging to a juvenile to this genus, comprising part of a lower jaw, two vertebrae and a rib. Like other mosasaurs, it probably preyed on the large ammonoids and occasional fish that abounded in the Late Cretaceous seas. *Moanasaurus* was a large mosasaur with a skull approximately 78 cm long, making the complete animal about 12 m long. Its stout teeth are broadly faceted with bulbous roots.

TECHNICAL DATA *Moanasaurus* is characterised by its robust, broad skull, which has a large, broad frontal bone, the skull being widest across the union of the postorbitofrontal and the parietal; there are 15 maxillary teeth; the basioccipital unit is solidly sutured with pro-otic, opisthotic exo-ccipital, supratemporal, supraoccipital and squa-mosal; atlas centra are fused to the axis; the humerus is short and massive (Wiffen 1980). Wright (1989) doubts the validity of *Moanasaurus* as a genus and, instead, regards it as a junior synonym of *Mosasaurus*. Wiffen (1990b) maintained that it is a distinct genus.

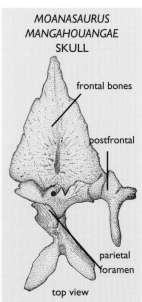

MOANASAURUS
MANGAHOUANGAE
SKULL

frontal bones

postfrontal

parietal
foramen

top view

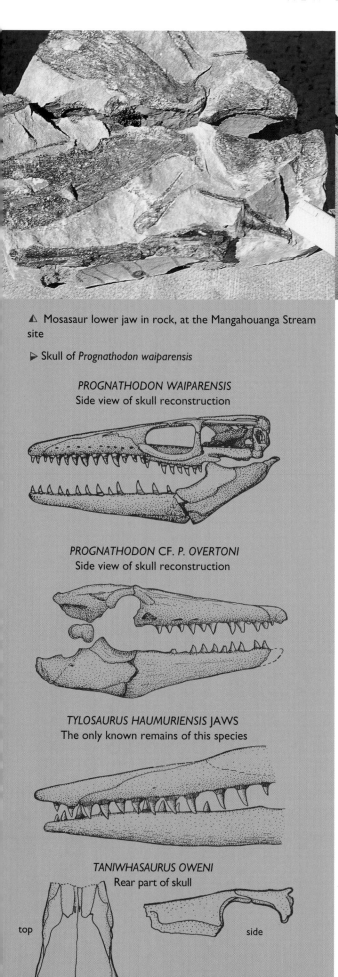

▲ Mosasaur lower jaw in rock, at the Mangahouanga Stream site

▷ Skull of *Prognathodon waiparensis*

PROGNATHODON WAIPARENSIS
Side view of skull reconstruction

PROGNATHODON CF. P. OVERTONI
Side view of skull reconstruction

TYLOSAURUS HAUMURIENSIS JAWS
The only known remains of this species

TANIWHASAURUS OWENI
Rear part of skull

top

side

(as first figured by Hector 1874)

GENUS PROGNATHODON

SPECIES *Prognathodon waiparensis* Welles and Gregg 1971

AGE Late Cretaceous (Maastrichtian)

LOCALITY Haumuri Bluff, Cheviot and Waipara district, South Island (Laidmore Formation)

SPECIES *Prognathodon* cf. *P. overtoni*

AGE Late Cretaceous

LOCALITY Mangahouanga Stream site, North Island (Maungataniwha Sandstone)

Prognathodon is a genus known from the Late Cretaceous of Europe, USA and New Zealand. The New Zealand species *P. waiparensis* takes its name from the locality, and is represented by a well-preserved complete skull more than 1 m in length, atlas-axis and some 14 vertebrae, part of the forelimb paddle and some ribs. The skull had lain on the sea floor for some time before it was buried, as indicated by its pitted appearance from gasteropod borings. Another nearly complete skull was found from the North Island Manga-houanga Stream site by Joan Wiffen and colleagues (Wiffen 1990a). It has close resemblances to the species from North America, *P. overtoni*, and so has been provisionally identified as this species, pending finds of further material.

TECHNICAL DATA *Prognathodon waiparensis* is characterised by its suprastapedial process of the quadrate being extremely constricted to less than half its distal breadth, by having a very large sulcus for the depressor mandibulae profundus, a massive infrastapedial process completely fused to the suprastapedial process muscle, and cervical pedicles that are short and set far forward (Welles and Gregg 1971). It has 11 teeth in each maxilla.

GENUS TYLOSAURUS

SPECIES *Tylosaurus haumuriensis* (Hector 1874)

AGE Late Cretaceous (Maastrichtian)

LOCALITY Haumuri Bluff, Cheviot and Waipara district, South Island (Laidmore Formation)

Tylosaurus is one of the largest and best known of all mosasaurs, having been recognised from Late Cretaceous deposits in North America, Europe, Canada and New Zealand. The New Zealand species, which is named after Haumuri Bluff, was enormous, with a lower jaw some 70 cm long, suggesting a total skull length of about 1.1 m long. This makes it the second-largest known species, next to the giant North American form *T. proriger*, whose skull is nearly 1.2 m long (Russell 1967). The total length of this animal can be estimated—from the known length of the skeleton of *T. proriger*—as being about 7 m. The lower jaw of this species has thin walls, the teeth having strong roots to strengthen the overall structure of the jaw.

TECHNICAL DATA *Tylosaurus haumuriensis* is characterised by its short, toothless ramus of the premaxilla that slants anterodorsally, yet is deep above the second tooth; the maxilla has 10–12 teeth, bearing five facets of the lateral surface, and the dentary of the lower jaw has a downwards curving anterior tip (Welles and Gregg 1971).

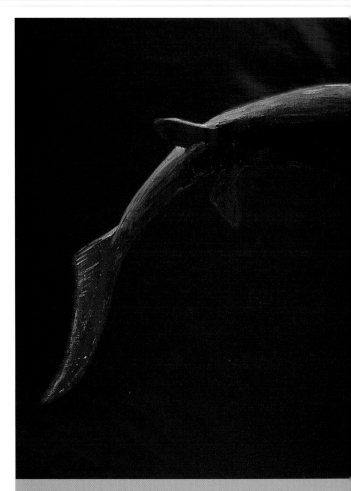

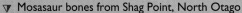

◁ Komodo dragon from Indonesia, the largest living lizard, is a close relative of the extinct mosasaurs

▽ Mosasaur bones from Shag Point, North Otago

▲ Reconstruction by Brian Choo of the slender mosasaur *Rikisaurus tehoensis*

NEW ZEALAND MOSASAUR SKULL

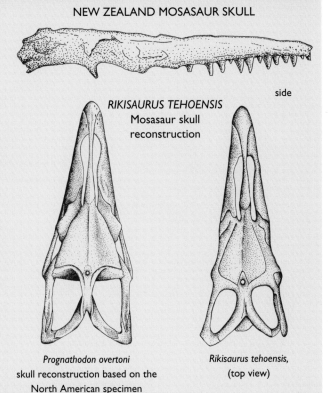

side

RIKISAURUS TEHOENSIS
Mosasaur skull
reconstruction

Prognathodon overtoni
skull reconstruction based on the
North American specimen
(top view)

Rikisaurus tehoensis,
(top view)

GENUS TANIWHASAURUS

SPECIES *Taniwhasaurus oweni* Hector 1874

AGE Late Cretaceous (Maastrichtian)

LOCALITY Haumuri Bluff, Cheviot and Waipara district, South Island (Laidmore Formation)

This mosasaur was one of the first endemic forms recognised from New Zealand, having been described by Hector in 1874 from a well-preserved partial skull. However, other material assigned to this genus by Hector has been relegated to indeterminate mosasaur status by Welles and Gregg (1971), so the comments here only pertain to the original skull. *Taniwhasaurus* is named after the Maori word 'Taniwha', for a large mythical dragon-like animal, and the Greek word 'sauros', meaning 'lizard'. It was a moderate size for a mosasaur, the frontal bones being 15 cm long, so the overall skull would be about half a metre long. The teeth have a cutting ridge along their biting margins.

TECHNICAL DATA *Taniwhasaurus* is characterised by, amongst other features, its skull having a post-frontal bone that overlaps far forwards on the pre-frontal; by a very broad frontal, which forms a median dorsal ridge, and has a broad posterolateral lappet; by a prefrontal that is large and forms part of the narial border; by the prefrontal and post-frontal just meeting below the frontal, with the prefrontal extending well forwards of the posterior margin of the naris; and by very small pterygoid teeth (Welles and Gregg 1971).

GENUS RIKISAURUS

SPECIES *Rikisaurus tehoensis* Wiffen 1990

AGE Late Cretaceous (Campanian–Maastrichtian)

LOCALITY Mangahouanga Stream site, North Island (Maungataniwha Sandstone)

Rikisaurus is based on a well-preserved adult skull about half a metre long, which shows large teeth, plus about four vertebrae. *Rikisaurus* takes its name from a Maori word meaning 'small' and 'sauros' ('lizard'), and the species name is for the Te Hoe Valley, where the specimen was found. It was a slender, medium-sized mosasaur, probably reaching about 8 m total length, and is believed to be of intermediate organisation between *Clidastes* and *Mosasaurus* from North America.

TECHNICAL DATA *Rikisaurus* is defined as a small mosasaurine with a long, tapering muzzle, V-shaped premaxilla, 13–14 teeth in the maxilla, and 9 in the pterygoids. All the teeth are small,

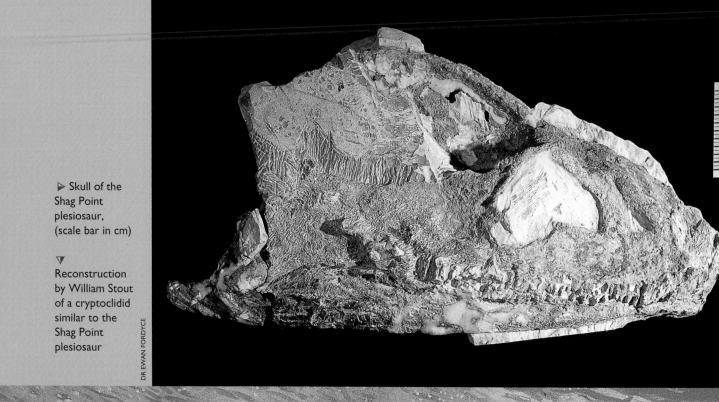

▷ Skull of the Shag Point plesiosaur, (scale bar in cm)

▽ Reconstruction by William Stout of a cryptoclidid similar to the Shag Point plesiosaur

DR EWAN FORDYCE

bicarinate with smooth enamel surfaces. The mandible is slender, with 15–16 teeth. The lower jaws deepen midway, with a distinct semicircular coronoid, and also have an elongate posterior development.

GENUS INDETERMINATE

SPECIES Indeterminate

AGE Late Cretaceous (Maastrichtian)

LOCALITY Shag Point, North Otago, South Island (Katiki Formation)

Isolated bones of a mosasaur are known from the Shag Point locality. These comprise a series of associated vertebrae and other indeterminate bits of bone, and were collected by Dr Ewan Fordyce. The specimens are currently held in the collections of the Otago University Geology Department.

SUPERFAMILY PLESIOSAUROIDEA

FAMILY PLIOSAURIDAE

GENUS INDETERMINATE

SPECIES Indeterminate

AGE Late Cretaceous (Maastrichtian)

LOCALITY Waipara River and Haumuri Bluff district, South Island (Laidmore Formation)

Pliosaur remains were first recorded from New Zealand by Hector (1874), when he described 'Polycotylus tenuis'. Welles and Gregg (1971) point out that these remains were described from undiagnostic bones, so the species name is of dubious value. Other material described as 'Plesiosaurus australis' by Owen in 1861 was also recognised by Welles and Gregg as belonging to a pliosaur, but because it is only known from some neck vertebrae, there is not enough material to warrant its being a new form. Thus, we have only tantalising fragments and isolated bones of pliosaurids from the South Island sites, none of which are currently able to be accurately identified as to genus.

▲ The Shag Point plesiosaur, a close relative of *Cryptoclidus*

▼ Excavating the plesiosaur at Shag Point, North Otago, in 1983

DR EWAN FORDYCE

DR EWAN FORDYCE

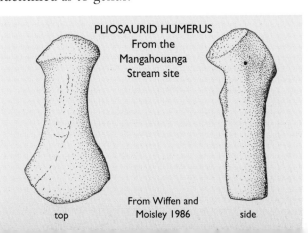

PLIOSAURID HUMERUS
From the
Mangahouanga
Stream site

top   From Wiffen and   side
           Moisley 1986

▲ Partial skeleton of a juvenile of the elasmosaur *Mauisaurus haasti*

▼ Forelimb paddle of the elasmosaur *Mauisaurus haasti* (Lectotype specimen)

Detail of the lag horizon containing phosphate nodules and fossil bones, exposed in a boulder at Mangahouanga Stream site

TUARANGISAURUS KEYSI

Skull in side view
(37 cm long)
Drawing by Danielle West

GENUS INDETERMINATE

SPECIES Indeterminate

AGE Late Cretaceous (Campanian–Maastrichtian)

LOCALITY Mangahouanga Stream site, North Island
(Maungataniwha Sandstone)

Pliosaurs were first recorded from the famous Mangahouanga Stream site in the North Island only in 1986 by Wiffen and Moisley, who, over ten years of collecting, found the remains of eight partial pliosaur skeletons. One of these shows the left underside of the shoulder girdle (coracoid bone) well-preserved, indicating it is different from the known pliosaurids. Other indeterminate pliosaurid remains from the same site include neck vertebrae and limb bones (propodials). One of these propodial bones measures nearly 34 cm in length, and has resemblances to that of *Polycotylus dolichops* (Wiffen and Moisley 1986). Other material includes a series of nine small vertebrae from a juvenile, and part of the pubis bone, as well as some other isolated vertebrae. Wiffen and Moisley (1986) suggested that all of this if one species, might represent a new genus and species of pliosaur.

FAMILY CRYPTOCLIDIDAE

Cryptoclidids were mostly Jurassic forms under 5 m long, intermediate between the short-necked pliosaurids and the long-necked elasmosaurids.

GENUS INDETERMINATE (new, as yet undescribed)

SPECIES Indeterminate

AGE Late Cretaceous (Maastrichtian)

LOCALITY Shag Point, North Otago, South Island
(Katiki Formation)

A large, articulated skeleton, mostly complete, of a plesiosaurian was collected in a boulder from the beach at Shag Point by Dr Ewan Fordyce, and took several years of painstaking preparation before its spectacular detail was revealed. Although the specimen is currently under study, it has been illustrated by Fordyce (1991). The Shag Point beast is a highly derived cryptoclidid, in having a much longer neck (with more vertebrae) than a normal cryptoclidid and yet retaining a regular neck vertebral morphology for the group. Its relationships are currently being investigated, although it does appear to be allied to the British *Cryptoclidus* and *Colymbosaurus*. Reconstructed, it would have measured approximately 7 m (Arthur Cruickshank, Ewan Fordyce, pers. comm. 1997).

FAMILY ELASMOSAURIDAE

GENUS MAUISAURUS

SPECIES *Mauisaurus haasti* Hector 1874

AGE Late Cretaceous (Maastrichtian)

LOCALITY Jed River, Gore Bay, 1 km from mouth, Waipara River district, South Island (Conway Siltstone Formation)

This beast is based on a pelvis and paddle illustrated by Hector in 1874, and several other bones, including vertebrae, limb bones, other paddle bones, parts of the shoulder girdle and pelvic girdle, subsequently discovered and prepared; but no skull material has yet been found. It was a large, long-necked plesiosaur, about 12 m maximum length, making it the largest long-necked elasmosaur yet found in the Southern Hemisphere. The name comes from 'Maui', a figure in Maori mythology who is credited with, among other things, raising the New Zealand landmass from the sea.

TECHNICAL DATA *Mauisaurus* is characterised by being an elasmosaur with dorsal centrum indices 80,117 : 145 and a sacral centrum 66,103 : 132; the anterior border of the pubis is convex, the mesial border extending posteriorly nearly to the level of the posterior border of the ischial facet, the lateral border being sharply concave. Femur breadth is 67 per cent of length (343 mm), with a large hemispherical capitulum, and nearly separate trochanter, dipping 30° posteriorly, with strong muscular rugosities. The humerus has horizontal tuberosity, broadly confluent with capitulum.

GENUS TUARANGISAURUS Wiffen and Moisley 1986

SPECIES *Tuarangisaurus keysi* Wiffen and Moisley 1986

AGE Late Cretaceous (Campanian–Maastrichtian)

LOCALITY Mangahouanga Stream site, North Island (Maungataniwha Sandstone)

This long-necked plesiosaur is known from a beautifully preserved skull found in one large boulder, plus a series of vertebrae with the atlas and axis bones, in another nearby boulder, as well as bits of skull in a smaller rock. They were found by Joan Wiffen and her husband, along with W. Moisley on 29 March 1978. It took some ten years of collecting before enough material was gathered to describe this beast and other marine reptiles from the same site. *Tuarangisaurus* takes its name from the Maori word meaning 'ancient'

and the Greek word for 'lizard'; the species name honours palaeontologist Ian Keyes. The skull of *Tuarangisaurus* has large front teeth that stick out between each other when the jaws are closed, making it an effective design for catching fishes.

TECHNICAL DATA An elasmosaur with a skull about 37 cm long, fused premaxillaries, a low central ridge, and five large teeth; the maxillaries have about fifteen teeth, and sutures with the premaxillary after the 5th tooth alveolus; there are 19–20 dentary teeth present, and a short mandibular symphysis, extending back between the 2nd and 3rd alveoli; the beak index is short;

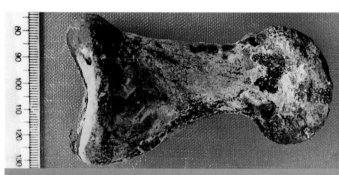

▲ Toe bone (pedal phalange) from a relatively large theropod dinosau

no parietal foramen is present; atlas-axis bones are fused. Other isolated bones that are attributed to *Tuarangisaurus keyesi* by Wiffen and Moisley (1986) include well-preserved pectoral girdle bones, humerus, complete front paddle bones, isolated teeth and many vertebrae.

ORDER ICHTHYOSAURIA

FAMILY ?STENOPTERYGIIDAE

Although the few, scant remains of Cretaceous ichthyosaurs from New Zealand are indeterminate as to genus and species, it is most probable that they belong in this family.

GENUS INDETERMINATE

SPECIES Indeterminate

AGE Early Cretaceous (Motuan, upper Albian)

LOCALITY Tinui district, southern part of the North Island (Makirikiri Formation)

Fleming *et al.* (1971) recorded three specimens of ichthyosaurs from three localities in the Tinui district, and these were tentatively referred to *Platypterygius* as this is the only currently known Cretaceous genus of ichthyosaur (McGowan 1972).

DINOSAURS

All of the following dinosaurs and pterosaurs are indeterminate as to genus and species, and come from the Late Cretaceous (Campanian–Maastrichtian) Mangahouanga Stream site, North Island.

ORDER SAURISCHIA
SUBORDER THEROPODA

Two bones, most likely from theropod dinosaurs, come from the Mangahouanga Stream site. The first dinosaur bone recognised from New Zealand was a small tail vertebra described by Molnar (1981), and identified as a possible theropod bone, although its affinity with ornithopods could not be dismissed. The second specimen is the only convincing evidence for large meat-eating dinosaurs roaming around the Late Cretaceous of New Zealand. It is a foot bone,

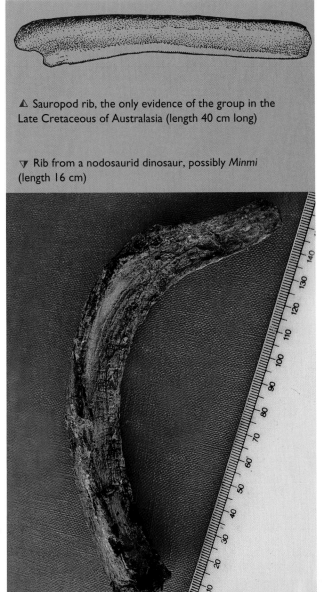

△ Sauropod rib, the only evidence of the group in the Late Cretaceous of Australasia (length 40 cm long)

▽ Rib from a nodosaurid dinosaur, possibly *Minmi* (length 16 cm)

JOAN WIFFEN

from the toes, measuring 97 mm long by 46 mm high. It is comparable in size to *Allosaurus fragilis*, a predator from the Late Jurassic of North America which was 12 m long.

TECHNICAL DATA The toe bone is identified as most probably from the first phalanx of pedal digit III from a large theropod because of its large size, its hollow shaft, and deep distal pit for the collateral ligament. It was illustrated and described by Molnar and Wiffen (1994).

SUBORDER SAUROPODOMORPHA

A single, slender bone fragment, measuring 40 cm long by about 6 cm wide, comes from a rib that would have been about a metre long, from a sauropod dinosaur in the size range of about 10–12 m. It is the only record of the sauropod dinosaurs from New Zealand, yet it is too incomplete to warrant any detailed comparisons with the major sauropod families.

TECHNICAL DATA The rib fragment is trapezoidal in distal cross-section and L-shaped at its proximal end, developing a flange-like shelf along its convex outer margin. The bone has coarse spongiosa inside, and most likely comes from the proximal end, near the head of the rib. It is mainly attributed to a sauropod because of its large size. It was illustrated and described by Molnar and Wiffen (1994).

ORDER ORNITHISCHIA
SUBORDER ANKYLOSAURIA

FAMILY ?NODOSAURIDAE

The known remains of armoured ankylosaurs from New Zealand consist of a well-preserved rib showing the head where it connected to the backbone, and two small vertebrae from the tail region. They indicate that the ankylosaur was about 2.5–3 m in length. The tail vertebrae are from the end of the tail, and indicate that the animal lacked a tail club, so was probably a nodosaurid rather than a true ankylosaur.

TECHNICAL DATA The rib is 16 cm long and is from the anterior of the chest. Its T-shaped cross-section is typical of thyreophoran ribs. The inclination of the head of the rib to its shaft, and its overall shape and proportions compare well to the *Minmi* sp., the almost complete specimen from Richmond, Queensland (see chapter 6).

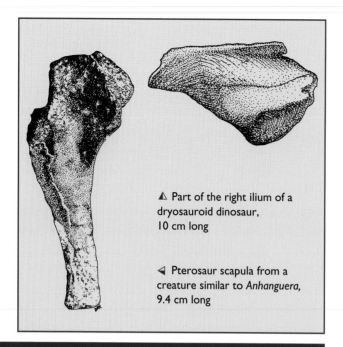

▲ Part of the right ilium of a dryosauroid dinosaur, 10 cm long

◄ Pterosaur scapula from a creature similar to *Anhanguera*, 9.4 cm long

SUBORDER ORNITHOPODA

The only fossil remains of an ornithopod dinosaur from New Zealand is part of a hip bone (right ilium) of a dryosauroid-like creature. The bone was described by Wiffen and Molnar (1988). In addition to this, a small toe bone, which is from either a theropod or an ornithopod, was described and illustrated by Molnar and Wiffen (1994).

TECHNICAL DATA The ilium fragment measures about 10 cm long, and comes from a creature that would have been of similar size to *Dryosaurus*—about 3 m long. The pedal phalanx measures about 41 mm long, it has few characteristic features, and could thus represent either an ornithopod like *Thescelosaurus* or a small theropod. Recently Mr Joseph McKee informed me that a new ornithopod bone has been found from the same site. It is currently under study.

PTEROSAURS

SUBORDER PTERODACTYLOIDEA

FAMILY ?ANHANGUERIDAE

A well-preserved impression of a pterodactyl scapula was found by Joan Wiffen inside a nodule from the Maungataniwha member (Tahora Formation). It measures 9.4 cm long, and is most likely from a member of the family Anhangueridae, which includes forms that have a well-developed keel on the front of the long snout. *Anhanguera* had a wingspan of about 4 m and lived in the Early

Cretaceous of South America, and also occurs in Queensland, Australia (see chapter 7).

TECHNICAL DATA The bone has two interpretations. Joan Wiffen believes it to be complete, save for the eroded surface, whereas Ralph Molnar believes the region around the glenoid is missing as it differs significantly from the known morphology of pterosaur scapulae. The presence of a cristiform posterior process would allow it to be placed with the anhanguerid pterodactyloids (Molnar and Wiffen 1994).

FAMILY INDETERMINATE

The partial arm bone (ulna) shows enough features to be identified as a pterosaur, although its exact affinities remain unclear. It closely resembles the same bone of *Santadactylus araripensis*, although lacks enough diagnostic features to be confidently placed in the same family. Attached to the matrix with the bone was the remains of a small, strongly curved tooth, indicating, if from the same individual, it was one of the toothed pterosaur families. The size of the bone indicates it came from a pterosaur with a wingspan of about 3.75 m.

TECHNICAL DATA The bone is the distal end of the left ulna. Its shaft measures 25 mm in diameter with bone 1.4 mm thick at its maximum. It has a prominent but short ventral ridge, which anteriorly has a flat, radial surface. Dorsally it has a sharp crest, placed presumably just near the dorsal margin of the radius. Proximal to the well-developed tuberculum is a small foramen. It has a small fovea carpalis 6.2 mm in diameter. The bone was described and illustrated by Wiffen and Molnar (1988).

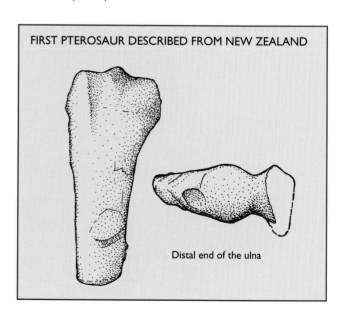

FIRST PTEROSAUR DESCRIBED FROM NEW ZEALAND

Distal end of the ulna

GLOSSARY of SCIENTIFIC TERMS

Words in brackets are the vernacular term derived from the formal scientific term, or alternatives in common usage.

acetabular pertaining to the hip joint (or acetabulum) of the leg

Actinopterygyii (actinopterygian) one of the three major subclasses of bony fishes (osteichthyans), containing the vast majority of all fishes living today. This subclass is characterised principally by having fins supported mainly by lepidotrichia; it also has ganoine layers in scales (examples: trout, goldfish, salmon)

alveolus hole in bone; especially holes for roots of teeth in jaws (plural: alveoli)

Amniota (amniote) group of tetrapods which lay hard-shelled eggs (amniote egg), or groups that have evolved from them—reptiles, birds, dinosaurs, mammals

amphibian group of tetrapods that reproduce by laying soft-shelled eggs in water (represented today by frogs, newts, salamanders and caecelians), of which there were a great many representatives in the Palaeozoic and Mesozoic Eras (for example, labyrinthodonts)

anapsid reptiles having a simple skull without any fenestrae (for example, turtles)

angiosperm vascular plants with seeds in ovaries, which includes most modern groups of plants that reproduce by flowers (for example, eucalypts, banksias, daisies)

ankylosaur group of 'armoured' ornithischian dinosaurs that have dermal plates covering the back; advanced members have a club tail (family: Ankylosauridae)

anterior The front end or surface of an organism or object (the anterior end of a car is the front grille)

anthracosaur group of extinct amphibians that have the skull and cheek loosely attached, and a closed palate

Archaean a period of geological time from 4.5 billion years ago to 2.5 billion years ago (part of Precambrian time)

archosauromorph one of the two major groups of reptiles; it includes crocodilians, pterosaurs, thecodonts, dinosaurs and birds (also called archosaurs)

astragalus ankle bone that articulates with the lower leg bone (tibia). Astragalar process is the high dorsal process on the astragalus of some theropods

atlas–axis vertebrae The first (neck joint) vertebra, which articulates with the skull and the axis vertebra (the second neck vertebra)

benthic bottom-dwelling, that is, living near the floor or bed of the sea, river or lake

biostratigraphy using fossils to determine the relative ages of rocks by comparing them with assemblages of fossils from other regions where the rocks above and/or below the fossil horizons are well-dated by radiometric methods; constraining the upper and lower possible age ranges for the fossil assemblage

biota total animal and plant numbers in a region

bipedal walking upright on the hind limbs

brachyopid one of the groups of Mesozoic labyrinthodont amphibians; it has a short head with large eyes (family: Brachyopidae)

calcaneum one of the principal ankle bones, adjacent to the astragalus, sometimes fused to it. The calcaneal notch is the notch on the calcaneum for the astragalus

calcification process of fossilisation whereby calcite is deposited in pore spaces and cavities of bones

calcite mineral formed of calcium carbonate, the principal component in limestone

carnosaur large tetanuran theropods placed within the clade Carnosauria (e.g. *Acrocanthosaurus, Allosaurus, Giganotosaurus*); traditionally used to describe any large theropod

Carnosauria clade containing tetanuran theropods more closely related to *Allosaurus* than to coelurosaurs/birds; typically large predators with prominent nostrils and cranial ornamentation.

Cenozoic Era the most recent of the three main eras of geological time, from 65 million years ago to the present (means 'modern life'; sometimes alternatively spelt Cainozoic)

centrum the main, spool-shaped body of a vertebra (plural: centra); the neural arch sits atop the centrum

ceratopsians a group of plant-eating dinosaurs characterised by their frilled head, often bearing horns (for example, *Triceratops*)

Ceratosauria group of theropod dinosaurs with tails stiffened for half of their length and usually four fingers per forearm; includes abelisaurids, ceratosaurids and coelophysids

clades related groups of organisms; those sharing specialised or derived evolutionary characteristics

cladogram diagram showing clades of related organisms that reflect the distribution of their derived character states

coelurosaur any member of the clade Coelurosauria; traditionally used to describe any small theropod dinosaur, *Coelurophysis* (family: Coelurosauridae)

Coelurosauria clade comprising tetanuran theropods which share a suite of bird-like features not found in carnosaurs; includes tyrannosaurs, dromaeosaurids, ornithomimids and others; traditionally the Coelurosauria included all small theropods, some of which (e.g. *Coelophysis*) are now classified as ceratosaurians

condyle knobby end process on the end of a limb bone for movable articulation with next bone—for example, medial condyle, lateral condyle on end of tibia

continental crust the top or 'outer skin' layer of the Earth, 5–150 km thick, which sits atop the mantle

continental drift the movement of continental masses on the earth. *see* Plate Tectonics

convergence *see* parallel evolution

coprolite   fossil dung

coracoid   one of the main bones in the shoulder of tetrapods

coronoid   bone (and coronoid process) of the lower jaw; it has an ascending process that goes up to the jaw joint in some forms

cosmopolitan   found all over the world; having a widespread distribution

cranium (cranial)   pertaining to the skull or head bones of an animal

Cretaceous   geological time period from 135 to 65 million years ago

crista   ridge

crista obliqua   special ridge on the tooth of mammals

Crossopterygyii (crossopterygian)   subclass of lobe-finned predatory osteichthyan fishes characterised by their kinetic skulls (for example, *Latimeria, Eusthenopteron*)

crust   outermost layer of the Earth, which can be up to 150 km thick over land (continental crust), and about 5 km thick underlaying the ocean (oceanic crust)

cryptodire   turtles that can retract the neck into the shell by flexure of the side of their necks (suborder: Cryptodira)

cultriform process   knife-like, main anterior process of the parasphenoid in amphibians

deltopectoral crest   well-developed flange on the proximal head of the humerus in dinosaurs, which is particularly well-developed in theropods—for the attachment of deltoid and pectoralis muscles

dentary   lower jaw bone, bearing a tooth row; this is the main bone forming the lower jaw in reptiles, and the only bone in the lower jaw of mammals

dentine   hard, dense tissue found in the teeth of tetrapods, also found in the teeth, spines and dermal bones of many primitive fishes; there are many kinds of dentine based on the arrangements of nutritive canals and orientation of fibres within the dentine layer

dermal bone   bone formed in the dermis of the skin, often with surface ornamentation; includes all the outside skull bones of a fish and many plate-like bones in skulls of tetrapods

diapsid   condition of reptilian skull having two openings (fenestrae) behind the orbit

dicynodonts   group of Triassic, mammal-like reptiles with large canine tusks

digit   single finger or toe. For digital nodes *see* phalangeal nodes

dinosaur   extinct group of reptiles that walked with the weight transferred to the feet through the metatarsal bones, and had diapsid-type skulls

distal   pertaining to part of an organism that is furthest away from the head—for instance, the hand is at the distal end of the arm

dorsal   top (the dorsal view of a car is that looking down on its roof); the dorsal surface of a bone is the surface seen from above when in life position

dryosaurid   hypsilophodont-like, ornithischian dinosaurs that lack teeth on the premaxilla

durophagous   feeding by crushing prey, particularly hard-shelled items; creatures that do this usually have hardened tooth-plates rather than sharp, pointed teeth

elasmosaur   one of the families of plesiosaurids; it has a very long neck, a small head, and about 72 cervical vertebrae (for example, *Woolungasaurus*)

elmisaurid   specialised group of small theropod dinosaurs that inhabited North America and Asia in the Late Cretaceous

endemic   in a biological gene confined to a certain geographic region; wombats are endemic to Australia, for instance, meaning that they are not found anywhere else in the world

endocast   internal mould of the brain cavity formed by sediment infilling a skull (or can be made using plaster or rubber)

endochondral bone   bone formed from a cartilage core, which replaces the cartilage frame (for example, the limb bones)

eukaryotes   organisms having a cell membrane, thus the DNA is divided from the rest of the cell (that is, in single-celled organisms, those having a nucleus)

eutherians   mammals that carry the foetus to full term and give birth to well-developed babies (that is, have a placenta inside the womb feeding the foetus); also called placental mammals (for example, sheep, cat, whale, mouse)

exoccipital   bones at the rear of the braincase in amphibians, reptiles and mammals, immediately lateral to the foramen magnum

femur   thigh bone in the leg (plural: femora)

fenestra   hole or opening within a bone, generally refers to skull

fibula   the minor of the two lower leg bones in tetrapods, sits next to the tibia

foramen magnum   The hole at the back of the skull for the spinal cord

fossil   remains, impressions or traces, or casts thereof, of previous life on this planet; usually it is the remains of something that has been buried and subjected to processes of fossilisation, thus in most cases (but not always) having had its appearance altered—for instance a bone that has been lithified as a result of impregnation by chemicals during burial

frontals   paired bones in the midline of the skull roof in osteichthyan fishes and tetrapods; the frontals enclose the pineal opening or foramen

genus   a taxonomic term for a group of species which have characteristic features that unite them as a distinct group— *Homo* is the genus of man, for example, uniting various species such as *Homo sapiens, Homo ergaster, Homo erectus* and *Homo habilis* (plural: genera)

glaciation   ice age. Glacial deposits are those sedimentary rocks built up as a result of the melting of glaciers (rivers of ice) and the dumping of their carried sediment load (moraine)

gnathostomes   vertebrate animals that have jaws (that is, most fishes, amphibians, reptiles, dinosaurs, birds, mammals)

Gondwana   the ancient southern supercontinent that included Australia, Antarctica, South America, South Africa, Arabia, and parts of the Middle East and Southeast Asia (means 'land of the Gonds'); the break up of Gondwana was most prevalent during the Mesozoic Era at the time of the dinosaurs (used to be called 'Gondwanaland')

gymnosperm   group of vascular plants that reproduce by producing cones, such as all conifers (technically, the ovaries are not enclosed in ovules)

hadrosaurs   a group of duck-billed, plant-eating dinosaurs, so far known only from North America and Asia in the Late Cretaceous

holotype the specimen designated at the time it was described to represent the characteristic features of that species (or subspecies)

horizons layers of sedimentary rock

humerus upper arm bone of any tetrapod or lobe-finned fish

hyomandibular large bone of the hyoid arch (first gill arch) that braces the jaw articulation in many primitive fish groups, lying against the palatoquadrate; in tetrapods it becomes modified as the stapes bone of the inner ear

hypantrum paired process at the front of the neural arch on a vertebra; articulates with hyposphene

hyposphene articulation pits on the posterior face of a vertebra, on the neural arch, for receiving the hypantrum of the vertebra posterior to it

hypsilophodontids small, fleet-footed, bipedal plant-eating ornithischian ornithopodan dinosaurs (for example, *Leaellynasaura*)

ichnogenus form genus characterised on footprint data only; a concept of an animal that made the tracks (also, ichnofamily, ichnotaxa, etc.)

ichthyosaur group of extinct marine reptiles that acquired dolphin-like body shapes, with well-developed dorsal fins and tail fins, and were generally long-snouted with many teeth for catching fish; they gave birth to live young

iguanodonts a group of ornithischian dinosaurs, large ornithopods, with a thumb spike

infradiapophysial buttress thin ridge of bone on the side of the neural arch (of a vertebra) which is in front of the cavity (infraprediapophysial fossa) below the main lateral process (zygapophysis)

interclavicle median ventral dermal bone of the shoulder girdle in tetrapods and some fishes; situated ventrally in between the clavicles

intercondylar groove groove between the condyles on the end of a limb bone

interpterygoid vacuities holes in the palate of amphibians, between the pterygoid bones

intramural cavity the vertebrae of some sauropods are almost hollow: the spaces are called intramural cavities and the thin strip of bone between these cavities is called the intramural wall

invertebrates animals lacking a bony skeleton (for example, arthropods, coelenterates)

jugal bone in the cheek of osteichthyan fishes and tetrapods

Jurassic a geological time period from 205 to 135 million years ago

labyrinthodont primitive fossil amphibians that have teeth with infolded enamel and dentine

lachrymal small bone in osteichthyan fishes and tetrapods that is the anterior-most bone of the cheek unit (also spelt 'lacrimal') it sits under the eye and in fishes carries the infraorbital sensory-line canal

lateral the externally facing side of an object or organism

Laurasia ancient, great Northern Hemisphere continent, comprising much of North America, Greenland, Asia, and part of Europe

lepidosauromorph one of the two major groups of reptiles; it includes modern lizards and snakes, and some related extinct forms

lophs bulbous or prominent parts of teeth, especially molars

limestone sedimentary rock composed essentially of calcium carbonate; it is mostly from marine environments, but rare freshwater or algal limestones do also occur

malleolus means 'little hammer'. Medial malleolus is the knobby process of the distal end of the tibia

mammal group of reptile-derived, warm-blooded tetrapods that have a single lower jaw bone (dentary), and a jaw joint formed where the squamosal meets the dentary; this group also has teeth that are well-differentiated into molars, premolars, canines and incisors

maniraptoran specialised group of theropod dinosaurs with elongated forearms and hands, amongst other features (for example, dromaeosaurids, troodontids, birds)

mantle the layer of the Earth below the crust; it is in liquid form and lies at a depth of between about 30 km (average depth) and 2900 km; the lower mantle is below 1100 km, where density increases significantly

manus the hand; may pertain to trackway impressions

marsupials mammals that give birth to their young at an early stage of development and nurture their young in a pouch; they are also characterised by unique dental and skeletal features

maxilla (maxillary) upper toothed jaw bone in fishes and tetrapods

mesosaurs a group of small Permian reptiles that were fully aquatic, and inhabited Gondwana regions (South Africa, South America)

Mesozoic Era a time when dinosaurs existed; it comprises three geological time periods (Triassic, Jurassic, Cretaceous), and was from 250 million years ago to 65 million years ago

metacarpals main hand bones leading to, and supporting the digit bones (fingers)

metacristid specialised ridge on the molar teeth of some mammals

metatarsals main foot bones leading to, and supporting the digit bones (toes)

metatarso-phalangeal nodes impressions seen in fossil footprints of the swollen region of tissue over the heel

microfossil a fossil which is of microscopic size, generally retrieved by using bulk extraction techniques; usually applies to invertebrates such as foraminiferans or radiolarians, although vertebrate microfossils include scales and teeth of fishes, or small reptilian and mammalian teeth

monophyletic any group of organisms that are united by sharing a unique set of features that evolved once within the group; this means that the group—whether a genus, family or higher taxonomic group—can be readily characterised by its specialised anatomical features

monotremes egg-laying mammals (for example, the platypus and echidna)

morphology study of shape; morphological differences are variations in shapes of certain bones or certain characteristics of the organism

mosasaur group of Late Cretaceous marine reptiles with long, powerful tails that propelled them through the water, and large heads with big teeth; they are closely allied to the living monitor lizards (varanids)

naris   external openings for the incurrent and excurrent nostrils of fishes (plural: nares)

nasal   bone in the snout of osteichthyan fishes and tetrapods, it flanks the median snout bones and is often notched for the nares

nasal bulla   expanded region of the snout over the nasal bones (as in *Muttaburrasaurus*)

neoceratopsian   group of ornithischian thyreophoran dinosaurs that are more advanced than the psittacosaurids in having well-developed frills; sometimes also with horns developed on the frill

neural arch   the dorsal spine above the centrum of the vertebra which encloses the hole for the spinal cord; it may have lateral wings and other complex processes for articulation with the adjacent vertebrae

neurocranium   braincase, which is generally separate from the dermal bones of the skull roof; in fishes this is the cartilaginous or bony block encasing the brain

niche   the role played by each organism within its community—for example, top line predator, benthic filter feeder—combined with its lifestyle (sedentary, for example, or active)

notochord   cartilaginous rod in primitive vertebrates and in embryonic higher vertebrates that forms the framework for the ossification of the backbone. The notochordal canal is the remant hole for the notochord in vertebral centra

occipital   pertaining to the neck region or back of the skull; or of a bone that sits in that region of the braincase.

occlusal   the biting or chewing surface of a tooth

olecranon   the proximal process on the ulna ('funny bone')

omnivorous   eats anything—in other words, a mixture of meat and plant materials (as humans do)

opisthocoelous   pertains to vertebrae that are concave at both anterior and posterior faces of the centrum

orbit   hole in the skull for the eye. Orbital notch may refer to part of the skull or cheek margin for the eye

ornithischian   one of the two major groups of dinosaurs; it is those with a pelvis that has a long, posteriorly projecting division on the pubis, with a short anterior division of that bone, and includes only plant-eating forms (ornithopods, ceratopsians, ankylosaurs, stegosaurs, pachycephalosaurs)

ornithomimosaur   ostrich-like dinosaurs, a group of fast-running theropods that generally have long necks, long legs, long arms, and a few or no teeth in a beaked mouth

ossicle   small, rounded or polygonal unit of dermal bone; generally set in the skin, or surrounding the eyes

osteoderms   *see* scutes

Osteolepiformes (osteolepiform)   a group of extinct lobe-finned (crossopterygian) fishes which gave rise to the first amphibians

otic notch   a notch in the rear of the skull roof of some fossil amphibians for the tympanic membrane of the ear; it may be open to the rear of the skull (as in *Parotosuchus*) or closed (as in *Paracyclotosaurus*)

outcrops   visible exposures of rock at the surface of the Earth

oviraptorosaur   small, highly advanced theropod dinosaurs with short, deep, toothless jaws, and long arms (for example, *Oviraptor*)

paedomorphosis   the retention of sub-adult ancestral traits in the descendent adult (a process giving rise to new species in a lineage)

palaeobiology   the study of fossils as once-living organisms, which involves reconstructing their biological processes

palaeobotany   the study of fossil plants

palaeogeography   the study of past continental positions and past environments, largely using information from rocks, fossils and magnetic data entrapped in minerals when rocks formed

palaeontology   the study of fossils

palaeozoology   the study of fossil animals

Pangaea   ancient supercontinent formed of the union between Gondwana and Eurasia, which included all main landmasses we know today; it began to break up in the Triassic Period

parallel evolution   evolution of similar characters separately (independently) in two or more lineages of common ancestry—for example, wings in bats and pterosaurs (also known as convergence)

paraphyletic group   a set of organisms which are not monophyletic, that is, they do not have the same common ancestry

parapophysis   on a single backbone or vertebra, the parapophysis is the anterior connecting area on the lateral margin of the centrum

parasphenoid   dermal median bone of the palate; it may be covered with small teeth in fishes

parietals   paired bones in the rear midline of the skull-roof in fishes and tetrapods

pelvic girdle   the series of bones connecting the lower limb (leg) to the axial skeleton. The main bones in reptiles, birds and mammals are collectively termed the 'pelvis' when fused together. Separately, they are known as the ilium, ischium and pubis bones. The legs articulate in the acetabulum (joint) of the pelvis

peramorphosis   development of traits beyond that of the ancestral adult (a process giving rise to new species)

periosteal   pertaining to the outer layered surface of a bone

pes   foot; may pertain to trackway impressions (adjective: pedal)

phalangeal nodes   impressions seen in fossil footprints of the swollen region of tissue between the joints of the toe digits (also called digital nodes)

phosphatic minerals   minerals having phosphate in them; bone is composed largely of the phosphatic mineral hydroxylapatite

phosphatisation   process of fossilisation in which pore spaces and cavities in a bone are filled in by phosphatic minerals

phylogeny   the evolution of a group; pertaining to evolutionary relationships (adjective: phylogenetic)

pineal foramen   the so-called 'third eye' in primitive vertebrates, a small opening in the skull roof between the frontals (for a light-sensitive organ); the pineal opening exists in many tetrapod groups

plate tectonics   the study of the movement of the continents over time, and the geological effects caused by such movements (for example, earthquakes, tsunamis); blocks of the Earth's crust are termed 'plates'

plesiosaur (plesiosaurian) a group of extinct marine reptiles (plesiosaurians) that propelled themselves through the water by large paddles; usually divided into those with relatively long necks (plesiosaurids) and those with shorter necks and longer heads (pliosaurids)

pleurocentrum the large, anterior (front) part of the vertebral disc in an amphibian or early reptile skeleton

pleurocoel a large hole in the side of a vertebra to reduce its weight

pliosaur short-necked plesiosaurian (family: Pliosauridae)

postacetabular pertaining to the hip region posterior to the acetabulum

posterior the back end or surface of an organism or object (the posterior end of a car is the surface showing the rear number plate)

posterolateral towards the rear and outer side

posteroventral towards the rear and downwards to the belly

postparietal paired (or singular, if fused) dermal bone of the skull behind the parietals

predentary a bone at the front of the lower jaw in ornithischian dinosaurs

prefrontal bone in the skull roof of some tetrapods in front of the frontal

prezygapophysis articulation area on the front top part of the neural arch of a vertebra (before the zygapophysis), for the neighbouring vertebral arch

process part of a bone that projects prominently

procoelous pertains to vertebrae that have a strongly convex anterior face and a strongly concave posterior face

prosauropod *see* sauropodomorphs

proximal pertaining to the part of an organism closest to the head (for example, the proximal end of the arm is near the shoulder)

pterosaur group of dinosaur-like flying reptiles (archosauromorphs) that lived in the Mesozoic Era (colloquially known as 'pterodactyl')

pterygoids large, paired dermal bones of the palate

quadrate one of the upper jaw bones that actually articulates with the lower jaw

quadratojugal one of the cheek bones in tetrapods (and osteichthyan fishes), which forms the posteroventral corner of the cheek unit

quadrupedal walking on all four limbs

radius one of the two forearm bones in tetrapods, with the ulna

ramus elongated flat section of the lower jaw; bar-like section of any bone

saurischian one of the two major groups of dinosaurs; it is those with a pelvis that have a long, anteriorly projecting division on the pubis, with a very short posterior division of that bone, and includes two main groups: the meat-eating theropods and the long-necked sauropodomorphs

sauropodomorphs a major groups of saurischian dinosaurs containing the prosauropods and the sauropods, the latter being the large, long-necked plant-eaters such as *Apatosaurus* and *Brachiosaurus*

scapula shoulder-blade bone

scapulocoracoid fused scapula and coracoid bones of the shoulder girdle

scutes bony platelets set in the dermis (skin)—for instance, the bony parts covering the back of a crocodile are called 'dermal scutes' or 'osteoderms'

sedimentary basin topographical region of the Earth where sedimentary rocks are accumulating, such as an off-shore continental slope margin

sedimentary rocks rocks made up of many small grains of sediment which are held together by a matrix or cement (for example, sandstone, limestone, mudstone)

serra saw-like edge on tooth (plural: serrae, or serrations)

shoulder girdle series of bones that connect the upper limb (arm) to the axial skeleton; in reptiles the principal bones are the scapula and coracoid (or fused together as a scapulocoracoid), although some, including birds, have a furcula and others, such as turtles, may have a scapulo-precoracoid

species members of an animal or plant population which can interbreed; fossil species are recognised principally by similar sets of morphological traits in their preserved remains

squamates a group of reptiles including the lizards and snakes and some extinct groups, for example, mosasaurs

squamosal large dermal bone in the posterior region of the cheek in fishes and most tetrapods

subduction a process whereby one of the Earth's crustal plates is thrust underneath another crustal plate, which causes melting of the underthrust sheet of crust

sulcus groove or well-defined depression on the surface of a bone (plural: sulci)

synapsid reptiles that have a single, low fenestra in the skull behind the orbit; they were represented by mammal-like reptiles, and eventually gave rise to the mammals (for example, *Thrinaxodon*)

tabular a posterolateral skull-roof bone in amphibians, useful for diagnosing different genera; some amphibians have tabular horns developed, others have simple tabular bones

talonid specialised cusp on the molariform teeth of some mammals; a talonid basin is the depression next to the talonid

taxon a unit within a classification—for example, a species can be a taxon, so is a genus or a family, which is called a 'taxonomic group' (plural: taxa)

Teleostei (teleosteans) the largest group, and most advanced of all ray-finned fishes, characterised principally by the presence of uroneural bones in the tail; it includes most living fishes (for example, salmon, trout, goldfish, tuna)

temnospondyl a grouping of fossil amphibians (labyrinthodonts) that have large crescent-shaped intercentra and smaller paired pleurocentra in the vertebrae

temporal bar vertical rod or sheet of bone separating the orbit from the post-temporal fenestra in the skull of some tetrapods

Tetanurae a group of theropod dinosaurs (also called tetanurae) with fully stiffened tails and usually three or two fingers per forearm; it includes carnosaurs and coelurosaurs

tetrapod four-legged vertebrate animal; the term includes all amphibians, reptiles, birds and mammals, even if some of these forms may have secondarily lost limbs (such as snakes and legless lizards)

thecodont   a group of advanced archosaurian reptiles that gave rise to the first dinosaurs; so named because their teeth are set in sockets (for example, *Chasmatosaurus*)

theropod   a group of saurischian dinosaurs that are bipedal and carnivorous; includes most large predatory forms (for example, *Allosaurus, Tyrannosaurus*) as well as many other medium-sized families (for example, ornithomimosaurs, oviraptorosaurs)

thyreophoran   group of ornithischian armoured dinosaurs, comprising the stegosaurs and ankylosaurs and near relatives

tibia   major shin bone in the leg of tetrapods

titanosaurids   a group of sauropod dinosaurs that survived into the Late Cretaceous; some had body armour (for example, *Saltasaurus*)

Triassic   geological time period from 250 to 205 million years ago

trigonid   specialised cusp on the molar teeth of some mammals

trochanter   narrow, elongated ridge developed on the proximal region of the femur for attachment of leg muscles

ulna   major forearm bone of tetrapods, which has the olecranon process

ventral   the underside, or belly surface, of an object, organism or bone

zygapophysial   pertaining to the complex articulation surfaces on the neural arch of a vertebra

REFERENCES

Archer, M.A. 1997. Messing with mammalian meanderings. *Geological Society of Australia Abstracts* 48:6.

Archer, M.A., Flannery, T. Ritchie, A. & Molnar, R. 1985. First Mesozoic mammal from Australia. *Nature* 318: 363–66.

Bartholomai, A. 1966a. Fossil footprints in Queensland. *Australian Natural History* 15: 147–50.

—— 1966b. The discovery of plesiosaurian remains in freshwater sediments in Queensland. *Australian Journal of Science* 28: 437.

—— 1979. New lizard-like reptiles from the Early Triassic of Queensland. *Alcheringa* 3: 225–34.

Bartholomai, A. & Molnar, R.E. 1981. *Muttaburrasaurus*, a new iguanodont (Ornithischia: Ornithopoda) dinosaur from the Lower Cretaceous of Queensland. *Memoirs of the Queensland Museum* 20: 319–49.

Bennett, S.C. & Long, J.A. 1991. A large pterodactyloid pterosaur from the late Cretaceous (late Maastrichtian) of Western Australia. *Records of the Western Australian Museum* 15: 435–44.

Benton, M.J. 1979. Ecological succession among Late Palaeozoic and Mesozoic tetrapods. *Palaeogeography, Palaeoclimatology, Palaeoecology* 26: 127–50.

—— 1985. Classification and phylogeny of the diapsid reptiles. *Zoological Journal of the Linnean Society* 84: 97–164.

Brown, D.S. 1981. The English Upper Jurassic Plesiosauroidea (Reptilia) and a review of the phylogeny and classification of the Plesiosauria. *Bulletin of the British Museum of Natural History (Geology)* 35: 253–347.

Camp, C.L. & Banks, M.R. 1978. A proterosuchian reptile from the Early Triassic of Tasmania. *Alcheringa* 2: 143–58.

Campbell, J.D. 1965. New Zealand Triassic saurians. *New Zealand Journal of Geology and Geophysics* 8: 505–09.

Campbell, K.S.W. & Bell, M. 1977. A primitive amphibian from the Late Devonian of New South Wales. *Alcheringa* 1: 369–81.

Carroll, R.L. 1987. *Vertebrate paleontology and evolution*. Freeman & Co., New York.

Chapman, F. 1919. New or little known fossils in the National Museum: part 24. On a fossil tortoise in ironstone from Carapook, near Casterton. *Proceedings of the Royal Society of Victoria* 32: 11–13.

Chernin, S. 1977. A new brachyopid, *Batrachosuchus concordi* sp. nov., from the Upper Luangwa Valley, Zambia with a redescription of *Batrachosuchus brownei* Broom, 1903. *Palaeontologia Africana* 20: 87–109.

Chiappe, L.M., Norell, M.A. & Clark, J.M. 1996. Phylogenetic position of *Mononykus* (Aves: Alvarezsauridae) from the Late Cretaceous of the Gobi Desert. *Memoirs of the Queensland Museum* 39: 557–82.

Chinsamy, A., Rich, T. & Rich, P. V. 1996. Bone histology of dinosaurs from Dinosaur Cove, Australia. *Journal of Vertebrate Paleontology, vol. 16 (supplement to number 3), abstracts of the 56th annual meeting, New York,* p. 28A.

Colbert, E.H. & Merrilees, D. 1967. Cretaceous dinosaur footprints from Western Australia. *Journal of the Royal Society of Western Australia* 50: 21–25.

Coombs, W.P. Jr. & Molnar, R.E. 1981. Sauropods (Reptilia, Saurischia) from the Cretaceous of Queensland. *Memoirs of the Queensland Museum* 20: 351–73.

Cosgriff, J.W. 1965. A new genus of Temnospondyli from the Triassic of Western Australia. *Journal of the Royal Society of Western Australia* 48: 65–90.

—— 1967. Triassic labyrinthodonts from New South Wales. *Australia and New Zealand Association for the Advancement of Science, Geology Section, January 1967,* pp. K4–K5.

—— 1968. *Blinasaurus*, a brachyopid genus from Western Australia and New South Wales. *Journal of the Royal Society of Western Australia* 52: 65–88.

—— 1972. *Parotosuchus wadei*, a new capitosaurid from New South Wales. *Journal of Paleontology* 46: 545–55.

—— 1973. *Notobrachyops picketti*, a brachyopid from the Ashfield Shale, Wianamatta Group, New South Wales. *Journal of Paleontology* 47: 1094–1101.

—— 1974. Lower Triassic Temnospondyli of Tasmania. *Geological Society of America Special Paper* 149: 1–134

Cosgriff, J.W. & Garbutt, N.K. 1972. *Erythrobatrachus noonkanbahensis*, a trematosaurid species from the Blina Shale. *Journal of the Royal Society of Western Australia* 55: 5–18.

Cosgriff, J.W. & Zawiskie, J.M. 1979. A new species of the Rhytidosteidae from the *Lystrosaurus* zone and a review of the Rhytidosteioidea. *Palaeontologia Africana* 22: 1–27.

Cruickshank, A.R., Fordyce, E. & Long, J.A. 1998. Recent developments in Australasian sauropterygian palaeontology. *Records of the Western Australian Museum Supplement* (CAVEPS Symposium volume).

Cruickshank, A.R. & Long, J.A. 1997a. *Leptocleidus clemai* sp. nov., a new species of pliosaurid reptile from the Early Cretaceous Birdrong Sandstone of Western Australia. *Records of the Western Australian Museum* 18.

—— 1997b. First record of dinosaur bone from the Early Cretaceous of Western Australia. *Records of the Western Australian Museum* 18.

Currie, P.J., Vickers-Rich, P. & Rich, T.H. 1996. Possible oviraptorosaur (Theropoda, Dinosauria) specimens from the Early Cretaceous Otway Group of Dinosaur Cove, Australia. *Alcheringa* 20: 73–79.

Dayton, L. 1991. Missing dinosaurs turn up in Australia. *New Scientist* 131: 14.

Dong, Z.-M. 1985. A middle Jurassic labyrinthodont (*Sinobrachyops platicephalus* gen. et sp. nov.) from Dashanpu, Zigong, Sichuan Province. *Vertebrata Palasiatica* 23: 301–06.

Douglas, J.G. 1969. The Mesozoic floras of Victoria. Parts 1 & 2. *Memoir of the Geological Survey of Victoria* 28.

Drinnan, A.N. & Chambers, T.C. 1986. Flora of the Lower Koonwarra fossil bed (Korumburra Group), South Gippsland, Victoria. *Association of Australasian Palaeontologists Memoir* 3: 1–77.

Etheridge, R. Jr. 1888. On additional evidence of the Genus *Ichthyosaurus* in the Mesozoic Rocks ('Rolling Downs Formation') of northeastern Australia. *Proceedings of the Linnean Society of New South Wales* 2: 405–09.

—— 1897. An Australian sauropterygian (*Cimoliasaurus*),

converted into precious opal. *Records of the Australian Museum* 3: 19–29.

—— 1904. A second sauropterygian converted into opal, from the Upper Cretaceous of White Cliffs, New South Wales. *Records of the Australian Museum* 5: 306–16.

—— 1917. Reptilian notes: Megalania prisca, Owen, and *Notiosaurus dentatus,* Owen; lacertilian dermal armour, opaliized remains from Lightning Ridge, Proceedings of Royal Society of Victoria 29: 127–33.

Flannery, T.F. & Rich, T.H. 1982. Dinosaur digging in Victoria. *Australian Natural History* 20: 195–98.

Flannery, T.F., Archer, M., Rich, T.M. & Jones, R. 1995. A new family of monotremes from the Cretaceous of Australia. *Nature* 377: 418–20

Fleming, C.A., Gregg, D.R. & Welles, S.P. 1971. New Zealand ichthyosaurs—a summary, including new records from the Cretaceous. *New Zealand Journal of Geology and Geophysics* 14: 734–41.

Fletcher, H.O. 1948. Footprints in the sands of time. *Australian Museum Magazine* 9: 247–51.

Fordyce, R. E. 1991. A new look at the fossil vertebrate record of New Zealand. In P.V. Rich, R.F. Baird, E. Thompson and J. Monaghan (eds), *Fossil Vertebrates of Australasia,* Pioneer Design Studio, Monash University, Melbourne, pp. 1191–1316.

Gaffney, E.S. 1981. A review of the fossil turtles of Australia. *American Museum Novitates* 2720: 1–38.

—— 1991. The fossil turtles of Australia. In P.V. Rich, R.F. Baird, E. Thompson and J. Monaghan (eds), *Fossil Vertebrates of Australasia,* Pioneer Design Studio, Monash University, Melbourne, pp. 704–20.

Galton, P.M. 1974. The ornithischian dinosaur *Hypsilophodon* from the Wealden of the Isle of Wight. *Bulletin of the British Museum of Natural History (Geology)* 25: 1–152.

Galton, P.M. & Cluver, M.A. 1976. *Anchisaurus capensis* (Broom) and a revision of the Anchisauridae (Reptilia, Saurischia). *Annals of the South African Museum* 69: 121–59.

Gibbons, A. 1996. New feathered fossil brings dinosaurs and birds closer. *Science* 274: 720–21.

Glauert, L. 1952. Dinosaur footprints near Broome. *The West Australian Naturalist* 3: 82–3.

Gross, J. D., Rich, T.H. & Vickers-Rich, P. 1993. Dinosaur bone infection. *National Geographic Research and Exploration* 9: 286–93.

Hamley, T. 1997. Biogeography of the procolophonids. *Geological Society of Australia Abstracts* 48: 34.

Hammer, W.R. & Hickerson, W. J. 1994. A crested theropod dinosaur from Antarctica. *Science* 264: 828–30.

Haubold, H. 1971. Ichnia Amphibiorum et Reptiliorum fossilum. In O. Kuhn (ed), *Handbuch der Palaeoherpetologie* Vol. 18.

Hector, J. 1874. On the fossil Reptilia of New Zealand. *Transactions of the Proceedings of the New Zealand Institute* 6: 333–58.

Hildebrand, A.R. 1997. The Chicxulub impact and its role in determining the K/T Boundary. *Abstracts of the Fermor Meeting, 18–19th February 1997, The Geological Society,* Burlington House, London: 19–20.

Hitchcock, E. 1844. Report on Ichnolithology, or fossil footmarks, with a description of several new species, and the coprolites of birds, from the valley of Connecticut River, and of a supposed footmark from the valley of Hudson River. *American Journal of Science* 47: 292–322.

—— 1848. An attempt to discriminate and describe the animals that made the fossil footmarks of the United States, and especially of New England. *Memoirs of the American Academy of Arts and Sciences, ser. 2,* 3: 129–256.

Holtz, T.R. Jr. 1994. The phylogenetic position of the Tyrannosauridae: implications for theropod systematics. *Journal of Paleontology* 68: 1100–117.

Howie, A.A. 1972a. A brachyopid labyrinthodont from the Lower Trias of Queensland. *Proceedings of the Linnean Society of New South Wales* 9: 268–77.

—— 1972b. On a Queensland labyrinthodont. In K.A. Joysey and T.S. Kemp (eds), *Studies in vertebrate evolution,* Oliver and Boyd, Edinburgh and London, pp. 50–64.

Huene, F. von. 1932. Die fossile Reptil-Ordnung Saurischia, ihre Entwicklung und Geschichte. *Monographs in Geologie und Paläontologie* 1: 1–361.

Hunt, A.P., Lockley, M.G., Lucas, S.G. & Meyer, C.A. 1994. The global sauropod fossil record. *Gaia* 10: 261–79.

Huxley, T.H. 1859. On some amphibian and reptile remains from South Africa and Australia. *Quarterly Journal of the Geological Society of London* 15: 642–58.

Jack, R.L. & Etheridge, R. Jr. 1892. *Geology and palaeontology of Australia and New Guinea.* Government Printer, Brisbane. 2 vols.

Jell, P.A. 1983. An Early Jurassic millipede from the Evergreen Formation in Queensland. *Alcheringa* 7: 195–200.

Jupp, R. & Warren, A.A. 1986. The mandibles of the Triassic temnospondyl amphibians. *Alcheringa* 10: 99–124.

Kurochkin, E.N. & Molnar, R.E. 1997. New Material of enantiornithine birds from the Early Cretaceous of Australia. *Archeringa* 21: 291–97.

de Laubenfels, M.W. 1956. Dinosaur extinction: one more hypothesis. *Journal of Paleontology* 30: 207–18.

Long, J.A. 1990. *Dinosaurs of Australia and other animals of the Mesozoic Era.* Reed Books, Balgowlah, NSW.

—— 1991. The long history of fossil fishes on the Australian continent. In P.V. Rich, R.F. Baird, E. Thompson and J. Monaghan (eds), *Fossil Vertebrates of Australasia,* Pioneer Design Studio, Monash University, Melbourne, pp. 337–428.

—— 1992. First dinosaur bones from Western Australia. *The Beagle, Records of the Northern Territory Museum of Arts and Sciences* 9: 21–28.

—— 1993. *Dinosaurs of Australia and other animals of the Triassic, Jurassic and Cretaceous Periods.* Reed Books, Balgowlah, NSW.

—— 1995a. *The Rise of Fishes—500 Million Years of Evolution.* University of New South Wales Press, Sydney, and Johns Hopkins University Press, Baltimore, USA.

—— 1995b. A theropod dinosaur bone from the Late Cretaceous Molecap Greensand, Western Australia. *Records of the Western Australian Museum* 17: 143–46.

Long, J.A. & Cruickshank, A.R. 1996. First record of an Early Cretaceous theropod dinosaur bone from Western Australia. *Records of the Western Australian Museum* 18: 219–22.

—— 1998. Further records of plesiosaurian reptiles from the Jurassic and Cretaceous Periods of Western Australia. *Records of the Western Australian Museum* 18.

Long, J.A. & Molnar, R.E. 1998. *Ozraptor,* gen. nov., a new Jurassic theropod dinosaur from Western Australia. *Records of the Western Australian Museum* 19.

Longman, H.A. 1915. On a giant turtle from the Queensland Lower Cretaceous. *Memoirs of the Queensland Museum* 3: 24–29.

—— 1922. An ichthyosaurian skull from Queensland. *Memoirs of the Queensland Museum* 7: 246–56.

—— 1924. A new gigantic marine reptile from the Queensland Cretaceous, *Kronosaurus queenslandicus* new genus and species. *Memoirs of the Queensland Museum* 8: 26–28.

—— 1926. A giant dinosaur from Durham Downs, Queensland. *Memoirs of the Queensland Museum* 8: 183–94.

—— 1927. The giant dinosaur *Rhoetosaurus brownei*. *Memoirs of the Queensland Museum* 9: 1–18.

—— 1930. *Kronosaurus queenslandicus*, a gigantic Cretaceous pliosaur. *Memoirs of the Queensland Museum* 10: 1–7.

—— 1932. Restoration of *Kronosaurus queenslandicus*. *Memoirs of the Queensland Museum* 10: 98.

—— 1933. A new dinosaur from the Queensland Cretaceous. *Memoirs of the Queensland Museum* 13: 133–44.

—— 1935. Palaeontological notes. *Memoirs of the Queensland Museum* 10: 236.

Longman, H. 1941. A Queensland fossil amphibian (*Austropelor*). *Memoirs of the Queensland Museum* 12: 29–32.

—— 1943. Further notes on Australian ichthyosaurs. *Memoirs of the Queensland Museum* 12: 101–104.

Lundelius, E. Jr. & Warne, S. St J. 1960. Mosasaur remains from the Upper Cretaceous of Western Australia. *Journal of Paleontology* 34: 1215–17.

McCoy, F. 1867. On the occurrence of *Ichthyosaurus* and *Plesiosaurus* in Australia. *Annals and Magazine of Natural History*, ser. 3, 19: 355–56.

—— 1868. On the discovery of Enaliosauria and other Cretaceous fossils in Australia. *Transactions of the Proceedings of the Royal Society of Victoria* 8: 41–42.

—— 1869. On the fossil eye and teeth of the *Ichthyosaurus Australis* (McCoy) from the Cretaceous formations of the source of the Flinders River; etc. *Transactions of the Proceedings of the Royal Society of Victoria* 9: 77–78.

McGowan, C. 1972. The systematics of Cretaceous ichthyosaurs with particular reference to the material from North America. *University of Wyoming Contributions in Geology* 11: 9–29.

McLaughlin, S., Haig, D.W., Backhouse, J., Holmes, M.A., Ellis, G., Long, J.A. & McNamara, K.J. 1995. Oldest Cretaceous sequence, Giralia Anticline, Carnavon Basin, Western Australia: late Hauterivian-Barremian. *Journal of Australian Geology and Geophysics* 15: 445–68.

Molnar, R.E. 1980a. An ankylosaur (Ornithischia: Reptilia) from the Lower Cretaceous of southern Queensland. *Memoirs of the Queensland Museum* 20: 77–87.

—— 1980b. Reflections on the Mesozoic of Australia. *Mesozoic Vertebrate Life* 1: 47–60.

—— 1980c. Procoelous crocodile from the Lower Cretaceous of Lightning Ridge, New South Wales. *Memoirs of the Queensland Museum* 20: 65–75.

—— 1981. A dinosaur from New Zealand. In P. Vella and M. Creswell (eds). *Gondwana Five*, A.A. Balkema, Rotterdam, pp. 91–96.

—— 1982a. Australian Mesozoic reptiles. In P.V. Rich and E. Thompson (eds), *The Fossil Vertebrate Record of Australasia*, Monash University off-set printing unit, Melbourne, pp. 170–225.

—— 1982b. A catalogue of fossil amphibians and reptiles in Queensland. *Memoirs of the Queensland Museum* 20: 613–33.

—— 1985. *Minmi paravertebra*. The *Minmi*, an armoured dinosaur. In P.V. Rich and G. Van Tets (eds), *Kadimakara: Extinct vertebrates of Australia*, Pioneer Design Studio, Lilydale, Victoria, pp. 172–76.

—— 1986. An enantiornithine bird from the Lower Cretaceous of Queensland, Australia. *Nature* 322: 736–38.

—— 1987. A pterosaur pelvis from western Queensland, Australia. *Alcheringa* 11: 87–94.

—— 1991. Fossil reptiles in Australia. In P.V. Rich. R.F. Baird, E. Thompson and J. Monaghan (eds), *Fossil Vertebrates of Australasia*, Pioneer Design Studio, Monash University, Melbourne, pp. 605–702.

—— 1994. *Minmi*, all tanked up and ready to grow. *Dinonews*, The Western Australian Museum, Perth 7: 7–9.

—— 1995. Possible convergence in the jaw mechanisms of ceratopians and *Muttaburrasaurus*. In A. Sun & Y. Wang (eds), *Sixth Symposium on Mesozoic Terrestrial Ecosystems and Biota, short papers*, China Ocean Press, Beijing, pp. 115–17.

—— 1996a. Observations on the Australian ornithopod dinosaur *Muttaburrasaurus*. *Memoirs of the Queensland Museum* 39: 639–52.

—— 1996b. Preliminary report on a new ankylosaur from the Early Cretaceous of Queensland, Australia. *Memoirs of the Queensland Museum* 39: 653–68.

Molnar, R.E., Flannery, T.F. & Rich, T.H. 1981. An allosaurid theropod dinosaur from the Early Cretaceous of Victoria, Australia. *Alcheringa* 5: 141–46.

—— 1985. Aussie *Allosaurus* after all. *Journal of Paleontology* 59: 1511–513.

Molnar, R.E. & Frey, E. 1987. The paravertebral elements of the Australian ankylosaur *Minmi* (Reptilia: Ornithischia, Cretaceous). *Neues Jahrbuch für Geologie und Paläontologie, Abhandlungen* 175: 19–37.

Molnar, R.E. & Galton, P.M. 1986. Hypsilophodontid dinosaurs from Lightning Ridge, New South Wales, Australia. *Geobios* 19: 231–39.

Molnar, R.E. & O'Reagan, M. 1989. Dinosaur extinctions. *Australian Natural History* 22: 560–70.

Molnar, R.E. & Pledge, N.S. 1980. A new theropod dinosaur from South Australia. *Alcheringa* 4: 281–87.

Molnar, R.E. & Thulborn, R.A., 1980. First pterosaur from Australia. *Nature* 288: 361–63.

Molnar, R.E. & Wiffen, J. 1994. Polar dinosaur faunas from New Zealand. *Cretaceous Research* 15: 689–706.

Molnar, R.E. & Willis, P.M.A. 1996. A neosuchian crocodile from the Queensland Cretaceous. *Journal of Vertebrate Paleontology*, vol. 16 (supplement to number 3), Abstracts of the 56th Annual Meeting, New York, p. 27A.

Murray, P.F. 1985. Ichthyosaurs from Cretaceous Mullman Beds near Darwin, Northern Territory. *The Beagle, Records of the Northern Territory Museum of Arts and Sciences* 2: 39–55.

—— 1987. Plesiosaurs from Albian aged Bathurst Island Formation siltstones near Darwin, Northern Territory, Australia. *The Beagle, Records of the Northern Territory Museum of Arts and Sciences* 4: 95–102.

Norell, M.A., Clark, J.M., Chiappe, L.M. & Demberelyin Dashzeveg' 1995. A nesting dinosaur. *Nature* 378: 774–76.

Norman, D. 1985. *The Illustrated Encyclopedia of Dinosaurs*. Hodder & Stoughton, Sydney.

—— 1990. Problematic theropoda: Coelurosaurs. In D.W. Weishampel, P. Dodson, P. and H. Osmolska (eds), *The Dinosauria*, University of California Press, Berkeley, pp. 280–305.

Owen, R. 1855. Description of a cranium of a labyrinthodont reptile *Brachyops laticeps*, from Mangali, Central India. *Quarterly Journal of the Geological Society of London* 11: 37–39.

—— 1882. On an extinct chelonian reptile (*Notochelys costata*, Owen) from Australia. *Quarterly Journal of the Geological Society of London*, 38: 178–83.

Persson, P.O. 1960. Lower Cretaceous plesiosaurians (Rept.) from Australia. *Lunds Universitets Arsskrift*, Avd. 2, Bd. 56, (12): 1–21.

—— 1963. A revision of the classification of the Plesiosauria with a synopsis of the stratigraphical and geographical distribution of the group. *Lunds Universitets Arsskrift*, Avd. 2, Bd. 59, (1): 1–60.

—— 1982. Elasmosaurid skull from the lower Cretaceous of Queensland (Reptilia: Sauropterygia). *Memoirs of the Queensland Museum* 20: 647–55.

Pledge, N. 1980. Vertabrate fossils of South Australia. From The South Australia Year Book 1980. D. Woolman, Government Printer, South Australia.

Retallack, G. J. 1996. Early Triassic therapsid footprints from the Sydney Basin, Australia. *Alcheringa* 20: 301–14.

Rich, P.V., Rich, T.H., Wagstaff, B.E., McEwan Mason, J., Douthitt, C.B., Gregory, R.T. & Felton, E.A. 1988. Evidence for low temperatures and biologic diversity in Cretaceous high latitudes of Australia. *Science* 242: 1403–406.

Rich, P.V. & Van Tets, G. 1982. Fossil birds of Australia and New Guinea: their biogeographic, phylogenetic and bio-stratigraphic input. In P.V. Rich and E. Thompson (eds), *The fossil vertebrate record of Australasia*, Monash University off-set printing unit, Melbourne, pp. 236–384.

Rich, T.H. 1996. Significance of polar dinosaurs in Gondwana. *Memoirs of the Queensland Museum* 39: 711–17.

—— 1994. Neoceratopsians and ornithomimosaurs: dinosaurs of Gondwana origin? *Research and Exploration* 10: 129–31.

Rich, T.H., Flannery, T.F. & Archer, M. 1989. A second Cretaceous mammalian specimen from Lightning Ridge, New South Wales. *Alcheringa* 13: 85–88.

Rich, T.H. & Rich, P.V. 1989. Polar dinosaurs and biotas of the Early Cretaceous of southeastern Australia. *National Geographic Research* 5: 15–53.

—— 1994. Neoceratopsians and ornithomimsaurs: dinosaurs of Gondwana origin? *Research and Exploration* 10:129–31.

Rich, T.H., Vickers-Rich, P., Constantine, A., Flannery, T.F., Kool, L. & Van Klaveren, N. 1997. A tribospheric mammal from the Mesozoic of Australia. *Science* 278: 1438–42.

Ritchie, A. 1988. Who should pay for Australia's past? *Australian Natural History* 22: 368–70.

Ritchie, A. 1990. Return of the great sea monsters. *Australian Natural History* 23: 538–45.

Romer, A.S. & Lewis, A.D. 1960. A mounted skeleton of the giant plesiosaur. *Kronosaurus*. *Breviora* 112: 1–15.

Russell, D.A. 1967. Systematics and morphology of American mosasaurs. *Bulletin of the Peabody Museum of Natural History* 23: 1–241.

Santa Luca, A.P. 1980. The postcranial skeleton of *Hetero-dontosaurus tucki* (Reptilia, Ornithischia) from the Stormberg of South Africa. *Annals of the South African Museum* 79: 159–211.

Seeley, H.G. 1891. On *Agrosaurus macgillvrayi*, a saurischian reptile from the NE-coast of Australia. *Quarterly Journal of the Geological Society of London* 47:164–65.

Sengupta, D.P. 1995. Chugitosaurid temnospondyls from the Late Triassic of India and a review of the Family Chugito-sauridae. *Palaeontology* 38: 313–39.

Sereno, P.C. 1986. Phylogeny of the bird-hipped dinosaurs. *National Geographic Research* 2: 234–56.

Stephens, W.J. 1887. On some additional labyrinthodont fossils from the Hawkesbury Sandstone of New South Wales. *Proceedings of the Linnean Society of New South Wales* 2: 1175–195.

Talent, J.A., Duncan, P.M. & Handley, P.L. 1966. Early Cretaceous feathers from Victoria. *Emu* 66: 81–86.

Teichert, C. & Matheson, R.S. 1944. Upper Cretaceous ichthyosaurian and plesiosaurian remains from Western Australia. *Australian Journal of Science* 6: 167–70.

Thulborn, R.A. 1979. A proterosuchian thecodont from the Rewan Formation of Queensland. *Memoirs of the Queensland Museum* 19: 331–55.

—— 1983. A mammal-like reptile from Australia. *Nature* 303: 330–31.

—— 1984. The avian relationships of *Archaeopteryx* and the origin of birds. *Zoological Journal of the Linnean Society of London* 82: 119–58.

—— 1985a. Birds as neotenous dinosaurs. *Hornibrook Symposium abstracts, Records of the Geological Survey of New Zealand* 9: 90–92.

—— 1985b. *Rhoetosaurus brownei*. The giant Queensland dinosaur. In P.V. Rich and G. Van Tets (eds), *Kadimakara: Extinct vertebrates of Australia*, Pioneer Design Studio, Lilydale, Victoria, pp. 166–71.

—— 1986. The Australian Triassic reptile *Tasmaniosaurus triassicus* (Thecodontia: Proterosuchia). *Journal of Vertebrate Paleontology* 6: 123–42.

—— 1990a. *Dinosaur Tracks*. Chapman and Hall, London.

—— 1990b. Mammal-like reptiles of Australia. *Memoirs of the Queensland Museum* 28: 169.

—— 1994. Ornithopod dinosaur tracks from the Lower Jurassic of Queensland. *Alcheringa* 18: 247–58.

Thulborn, R.A., Hamley, T. & Foulkes, P. 1994. Preliminary report on sauropod dinosaur tracks in the Broome Sandstone (Lower Cretaceous) of Western Australia. *Gaia* 10: 85–96.

Thulborn, R.A. & Wade, M., 1979. Dinosaur stampede in the Cretaceous of Queensland. *Lethaia* 12: 275–79.

—— 1984. Dinosaur trackways in the Winton Formation (Mid-Cretaceous) of Queensland. *Memoirs of the Queensland Museum* 21: 413–518.

Thulborn, R.A. & Warren, A.A. 1980. Early Jurassic plesiosaurs from Australia. *Nature* 285: 224–25.

Upchurch, P. 1994. Sauropod phylogeny and palaeoecology. *Gaia* 10: 249–60.

Vickers-Rich, P. 1996. Early Cretaceous polar tetrapods from the Great Southern Rift valley, southeastern Australia. *Memoirs of the Queensland Museum* 39: 719–24.

Vickers-Rich, P. & Rich, T.H. 1993. Australia's polar dinosaurs. *Scientific American* 269: 50–55.

Vickers-Rich, P., Rich, T. H., McNamara, G. & Milner, A. 1998. Is *Agrosaurus macgillirrayi* Australia's oldest dinosaur? *Records of the Western Australian Museum Supplement* 57.

Wade, M. 1984. *Platypterygius australis*, an Australian Cretaceous ichthyosaur. *Lethaia* 17: 99–113.

—— 1990. A review of the Australian Cretaceous longipinnate ichthyosaur *Platypterygius* (Ichthyosauria, Ichthyopterygia). *Memoirs of the Queensland Museum* 28: 115–52.

Waldman, M. 1970. A third specimen of a lower Cretaceous feather from Victoria, Australia. *Condor* 72: 377.

—— 1971. Fish from the freshwater Lower Cretaceous of Victoria, Australia, with comments on the palaeo-environment. *Special Papers in Palaeontology* 9: 1–124.

Warren, A.A. 1977. Jurassic labyrinthodont. *Nature* 265: 436–37.

—— 1980. *Parotosuchus* from the Early Triassic of Queensland and Australia. *Alcheringa* 4: 25–36.

—— 1981a. The lower jaw of the labyrinthodont family Brachyopidae. *Memoirs of the Queensland Museum* 20: 285–89.

—— 1981b. A horned member of the labyrinthodont superfamily Brachyopoidea from the Early Triassic of Queensland. *Alcheringa* 5: 273–88.

—— 1982. Australian fossil amphibians. In P.V. Rich and E. Thompson (eds), *The Fossil Vertebrate Record of Australasia*, Monash University off-set printing unit, Melbourne, pp. 146–57.

—— 1985a. Two long snouted temnospondyls (Amphibia, Labyrinthodontia) from the Triassic of Queensland. *Alcheringa* 9: 293–96.

—— 1985b. An Australian plagiosauroid. *Journal of Paleontology* 59: 236–41.

—— 1991. Australian fossil amphibians. In P.V. Rich, R.F. Baird, E. Thompson and J. Monaghan (eds), *Fossil Vertebrates of Australasia*, Pioneer Design Studio, Monash University, Melbourne, pp. 569–90.

Warren, A.A. & Black, T. 1985. A new rhytidosteid (Amphibia, Labyrinthodontia) from the Early Triassic Arcadia Formation of Queensland, Australia, and the relationships of Triassic temnospondyls. *Journal of Vertebrate Paleontology* 5: 303–27.

Warren, A.A. & Damiani, R.J. 1996. A new look at members of the Superfamily Brachyopoidea (Amphibia, Temnospondyli) from the Early Triassic of Queensland and a preliminary analysis of brachyopoid relationships. *Alcheringa* 20: 277–300.

Warren, A.A. & Damiani, R.J. Northwood, C. & Yates, A. 1997. Palaeobiogeography of Australian fossil amphibians. *Geological Society of Australia Abstracts* 48:78.

Warren, A.A. & Hutchinson, M.N. 1983. The last labyrinthodont? A new brachyopoid (Amphibia, Temnospondyli) from the Early Jurassic Evergreen Formation of Queensland, Australia. *Philosophical Transactions of the Royal Society of London* (B) 209: 1–73.

—— 1987. The skeleton of a new hornless rhytidosteid (Amphibia, Temnospondyli). *Alcheringa* 11: 291–302.

—— 1988. A new capitosaurid amphibian from the Early Triassic of Queensland, and the ontogeny of the capitosaur skull. *Palaeontology* 31: 857–76.

—— 1990a. *Lapillopsis*, a new genus of temnospondyl amphibians from the Early Triassic of Queensland. *Alcheringa* 14: 149–58.

—— 1990b. The young ones—small temnospondyls from the Arcadia Formation. *Memoirs of the Queensland Museum* 28: 103–106.

Warren, A.A., Kool, L., Cleeland, M., Rich, T.H. & Vickers-Rich, P. 1991. An Early Cretaceous labyrinthodont. *Alcheringa* 15: 327–32

Warren, A.A., Rich, T.H. & Vickers-Rich, P. 1997. The last labyrinthodonts? *Palaeontographica* 247A: 1–24.

Warren, A.A. & Schroeder, N. 1995. Changes in the capitosaur skull with growth: an extension of the growth series of *Parotosuchus aliaciae* (Amphibia, Temnospondyli) with comments on the otic area of capitosaurs. *Alcheringa* 19: 41–46.

Warren, J.W. 1969. A fossil chelonian of probably Lower Cretaceous age from Victoria, Australia. *Memoirs of the National Museum of Victoria* 29: 23–28.

Watson, D.M.S. 1956. The brachyopid labyrinthodonts. *Bulletin of the British Museum of Natural History (Geology)* 2: 315–91.

—— 1958. A new labyrinthodont (*Paracyclotosaurus*) from the Upper Trias of New South Wales. *Bulletin of the British Museum of Natural History (Geology)* 3: 233–63.

Welles, S.P. 1983. *Allosaurus* (Saurischia, Theropoda) not yet in Australia. *Journal of Paleontology* 57: 196.

Welles, S.P. & Gregg, D.R. 1971. Late Cretaceous marine reptiles of New Zealand. *Records of the Canterbury Museum* 9: 1–111.

Welles, S.P. & Long, R.A. 1974. The tarsus of theropod dinosaurs. *Annals of the South African Museum* 64: 191–218.

White, T.E. 1935. On the skull of *Kronosaurus queenslandicus* Longman. *Occasional papers of the Boston Society of Natural History* 8: 219–28.

Wiffen, J. 1980. *Moanasaurus*, a new genus of marine reptile (Family Mosasauridae) from the Upper Cretaceous of North Island. *New Zealand Journal of Geology and Geophysics* 23: 507–28.

—— 1981. The first Late Cretaceous turtles from New Zealand. *New Zealand Journal of Geology and Geophysics* 24: 293–99.

—— 1990a. New mosasaurs (Reptilia; Family Mosasauridae) from the Upper Cretaceous of North Island. *New Zealand Journal of Geology and Geophysics* 33: 67–85.

—— 1990b. *Moanasaurus mangahouangae* or *Mosasaurus mangahouangae*?. *New Zealand Journal of Geology and Geophysics* 33: 1.

—— 1996. Dinosaurian palaeobiology: A New Zealand perspective. *Memoirs of the Queensland Museum* 39(3): 725–31.

Wiffen, J. & Moisley, W.L. 1986. Late Cretaceous reptiles (Families Elasmosauridae and Pliosauridae) from the Mangahouanga Stream, North Island. *New Zealand Journal of Geology and Geophysics* 29: 205–52.

Wiffen, J. & Molnar, R.E. 1988. First pterosaur from New Zealand. *Alcheringa* 12: 53–59.

Woodward, A.S. 1906. A tooth of *Ceratodus* and a dinosaurian claw from the Lower Jurassic of Victoria, Australia. *Annals and Magazine of Natural History* ser. 7 (18): 1–3, also *Records of the Geological Survey of Victoria* 2: 135–37.

Wright, K.R. 1989. *Moanasaurus mangahouangae* Wiffen (Squamata: Mosasauridae). *Journal of Palaeontology* 63: 126–27.

Yates, A.M. 1996. New material of the small Temnospondyl *Lapillopsis* from the Early Triassic Arcadia Formation of Australia. *Journal of Vertebrate Palaeontology*, 16, supplement to no. 3, Abstracts, p. 74.

INDEX

Page numbers in *italics* refer to illustrations.